Geographic Uncertainty in Environmental Security

NATO Science for Peace and Security Series

This Series presents the results of scientific meetings supported under the NATO Programme: Science for Peace and Security (SPS).

The NATO SPS Programme supports meetings in the following Key Priority areas: (1) Defence Against Terrorism; (2) Countering other Threats to Security and (3) NATO, Partner and Mediterranean Dialogue Country Priorities. The types of meeting supported are generally "Advanced Study Institutes" and "Advanced Research Workshops". The NATO SPS Series collects together the results of these meetings. The meetings are co-organized by scientists from NATO countries and scientists from NATO's "Partner" or "Mediterranean Dialogue" countries. The observations and recommendations made at the meetings, as well as the contents of the volumes in the Series, reflect those of participants and contributors only; they should not necessarily be regarded as reflecting NATO views or policy.

Advanced Study Institutes (ASI) are high-level tutorial courses intended to convey the latest developments in a subject to an advanced-level audience

Advanced Research Workshops (ARW) are expert meetings where an intense but informal exchange of views at the frontiers of a subject aims at identifying directions for future action

Following a transformation of the programme in 2006 the Series has been re-named and re-organised. Recent volumes on topics not related to security, which result from meetings supported under the programme earlier, may be found in the NATO Science Series.

The Series is published by IOS Press, Amsterdam, and Springer, Dordrecht, in conjunction with the NATO Public Diplomacy Division.

Sub-Series

A.	Chemistry and Biology	Springer
B.	Physics and Biophysics	Springer
C.	Environmental Security	Springer
D.	Information and Communication Security	IOS Press
E.	Human and Societal Dynamics	IOS Press

http://www.nato.int/science
http://www.springer.com
http://www.iospress.nl

Series C: Environmental Security

Geographic Uncertainty in Environmental Security

edited by

Ashley Morris

DePaul University
Chicago, U.S.A.

and

Svitlana Kokhan

National Agricultural University of Ukraine
Kyiv, Ukraine

Springer

Published in cooperation with NATO Public Diplomacy Division

Proceedings of the NATO Advanced Research Workshop on
Fuzziness and Uncertainty in GIS for Environmental Security and Protection
Kyiv, Ukraine
June 28–July 1, 2006

A C.I.P. Catalogue record for this book is available from the Library of Congress.

ISBN 978-1-4020-6437-1 (PB)
ISBN 978-1-4020-6436-4 (HB)
ISBN 978-1-4020-6438-8 (e-book)

Published by Springer,
P.O. Box 17, 3300 AA Dordrecht, The Netherlands.

www.springer.com

Printed on acid-free paper

CONTENTS

Preface vii

1. Fuzzy Regions: Theory and Applications 1
J. Verstraete, A. Hallez, G. De Tré

2. Mapping the Ecotone with Fuzzy Sets 19
C. Arnot, P. Fisher

3. Issues and Challenges of Incorporating Fuzzy Sets in Ecological Modeling 33
V.B. Robinson

4. Reliability of Vegetation Community Information Derived using Decorana Ordination and Fuzzy *c*-means Clustering 53
L. Bastin, P.F. Fisher, M.C. Bacon, C.N.W. Arnot, M.J. Hughes

5. A Rough Set-based Approach to Handling Uncertainty in Geographic Data Classification 75
P. Jankowski

6. Fuzzy Models for Handling Uncertainty in the Integration of High Resolution Remotely Sensed Data and GIS 89
J. Schiewe, M. Ehlers

7. Incompleteness, Error, Approximation, and Uncertainty: an Ontological Approach to Data Quality 107
A.U. Frank

8. A Flexible Decision Support Approach to Model ill-defined Knowledge in GIS 133
G. Bordogna, M. Pagani, G. Pasi

9. Development of the Geoinformation System of the State Ecological Monitoring.. 153
V.B. Mokin

10. Mapping Type 2 Change in Fuzzy Land Cover 167
P. Fisher, C. Arnot

11. Indexing Implementation for Vague Spatial Regions with R-trees and Grid Files 187
F.E. Petry, R. Ladner, M. Somodevilla

12. Association Rule Mining using Fuzzy Spatial Data Cubes 201
N. Işik, A.Yazici

13. Interactive Objects Extraction from Remote Sensing Images 225
V. Bucha, S. Ablameyko

14. Classification of Remotely Sensed Data 239
S. Kokhan

15. Sustainability and Environmental Security Management Tools 249
A. Gorobets

16. Remote Sensing and GIS Application for Environmental Monitoring and Accidents Control in Ukraine 259
D.K. Mozgoviy, O.I. Parshyna, V.I. Voloshyn, Y.I. Bushuyev

17. ProDec – Emergency Procedure Based on Fuzzy Notions for Catchment Management 271
J.W. Owsiński, A. Ziółkowski

Index 285

PREFACE

On June 28 through July 1, 2006, a NATO advanced research workshop was held in Kyiv, Ukraine. This meeting of scholars from both NATO and NATO partner countries brought together the leading researchers in the field of Fuzziness and Uncertainty in GIS for Environmental Security and Protection.

The papers based upon the presentations at this meeting are included in this book. They were all quite good, some focusing on the use of fuzzy sets in geography, others focusing on explicit environmental concerns in Ukraine. What this book cannot show is the camaraderie and spirit of cooperation that permeated the atmosphere of this workshop. These researchers, many of whom had never met before, are now colleagues, and are continuing to collaborate on research to this day, and probably beyond. Scholars from various positions and countries have chosen to work together in the spirit of cooperation to make the ideas presented in this book come to fruition.

At press time, we do not know if Ukraine will become a NATO member or not. In any case, the editors wish to thank the NATO Science Committee for their funding, encouragement, and work in making this workshop a reality. We hope that the spirit of cooperation and intellectual curiosity that was so lively at the conference will encourage the reader in the perusal of this work.

On behalf of the National Agricultural University of Ukraine, DePaul University, and the organizing committee of the NATO Advanced Research Workshop, we would like to express our gratitude to the NATO Public Affairs Division for their contribution in organizing this NATO ARW, as well as their contribution in preparing this publication. Also this workshop could not have taken place, without the hard work and contributions of both the Provost of DePaul University, Helmut Epp; and our host, the rector of the National Agricultural University of Ukraine, Dmytro O. Melnychuk.

Ashley Morris and Svitlana Kokhan, Editors

FUZZY REGIONS: THEORY AND APPLICATIONS

JÖRG VERSTRAETE *
TELIN – Ghent University, Sint Pietersnieuwstraat 41, 9000 Ghent, Belgium; jorg.verstraete@telin.ugent.be

AXEL HALLEZ
TELIN – Ghent University, Sint Pietersnieuwstraat 41, 9000 Ghent, Belgium; axel.hallez@telin.ugent.be

GUY DE TRÉ
TELIN – Ghent University, Sint Pietersnieuwstraat 41, 9000 Ghent, Belgium; guy.detre@telin.ugent.be

Abstract. Traditionally, information in geographic information systems (GIS) is represented as crisp information. While for many applications, this is a good enough approximation of reality, some models would benefit from having the inherent imprecision or uncertainty incorporated in the model. In literature, several ideas and concepts to improve on the crisp models have been considered. In the past, we have presented different models to represent and to work with the concept of regions, defined using fuzzy set theory in GIS systems. For such fuzzy regions, a number of approaches already have been described in detail. In this paper, we will elaborate on a fuzzy set approach and practical implementations of the concept. Apart from the concept, two developed techniques (one based on triangulated networks, one based on bitmap models) are presented along with some of the operators. An overview of application fields is provided to illustrate where and how the techniques can be used.

Keywords: fuzzy regions, spatial fuzzy set theory, fuzzy spatial data types

1. Introduction

Crisp regions are commonly defined by their outline, which usually takes the form of a closed polyline (Figure 1). In reality, regions are not always

*To whom correspondence should be addressed.

A. Morris and S. Kokhan (eds.), Geographic Uncertainty in Environmental Security, 1–17.

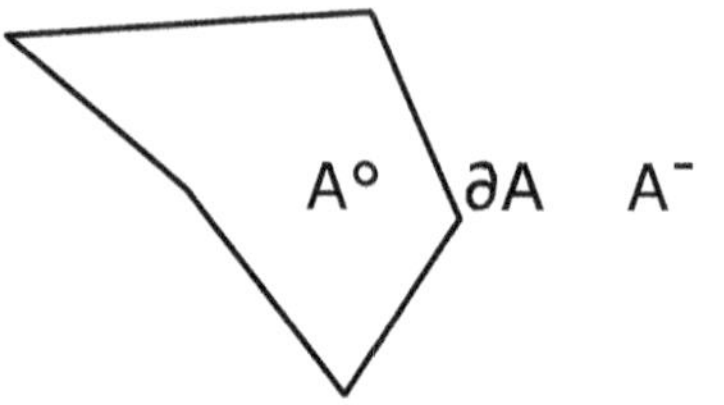

Figure 1. Representation of crisp regions

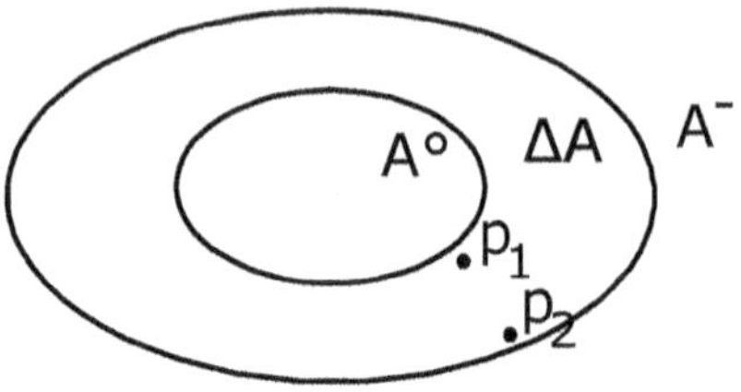

Figure 2. Representation of the broad boundary concept

crisp; consider the soil composition for example: the line where one type of soil stops and another starts cannot always be crisply defined. Also, regions that are the result of predictions (growth of cities for instance) can hardly be called crisp. The benefits of incorporating fuzziness in geographic databases are further explained in Morris (2001). In order to improve on the concept of crisp regions, several models have been developed to provide a richer model for regions in a geographic information system. Clementini (1994), Cohn and Gotts (1994) have presented similar models, based on changing the concept of one crisp boundary to two boundaries (an inner and an outer boundary, Figure 2). Clementini called it the *broad boundary* model, Cohn and Gotts refer to theirs as the *egg yolk* model.

While this is a natural way of expanding the concept of a region, it misses out on some aspects: neither of the models provides information for points that are inside the broad boundary (or to employ the egg-yolk terminology: in the egg, but not in the yolk). Basically, points are either inside the region, outside the region, or in the boundary: the points p_1 and p_2 on Figure 2 are treated as equals, yet for some applications it can be interesting to consider p_1 as belonging more to the A° than p_2. Furthermore, only topology has been considered in depth.

While it is possible to extend the boundary from two crisp boundaries to any number of crisp boundaries (as presented in Verstraete et al., 2000 and Hallez et al., 2002), it becomes cumbersome.

To improve on this, we have introduced the concept of fuzzy regions. A region is traditionally defined by an outline; all points (locations) inside this outline belong to the region. However, one can turn this around, and consider a region to be a set of points (locations).

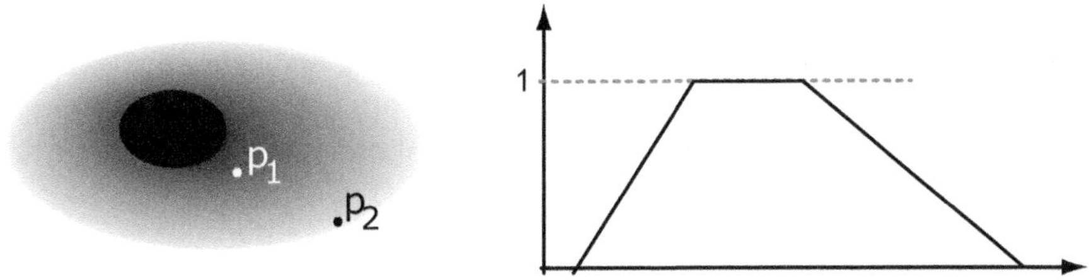

Figure 3. Concept of a fuzzy region, represented as a shaded two-dimensional region (left) and its cross section (right)

2. Fuzzy Regions

2.1. CONCEPT

Our concept of fuzzy regions does away with a predefined boundary. It is based on fuzzy set theory, and considers a region to be a fuzzy set of points (Figure 3).

In fuzzy set theory, fuzzy sets are defined as sets in which each element has a membership grade. This membership grade can have a multitude of interpretations (depending on the application) (Dubois and Prade, 1997): degree of membership, degree of uncertainty or degree of truth. These different interpretations can also be used for fuzzy spatial data types, but for the modelling of fuzzy regions only the first interpretation is suitable. The second interpretation is used in a similar approach to model fuzzy points, whereas the third interpretation can be useful to model query results.

For fuzzy regions, the concept of fuzzy sets is used in a two-dimensional space. A region will not be defined by a boundary, but by the points belonging to it; after all, even in a crisp GIS, a region can still be considered as a set of points. To achieve this, a fuzzy region will be defined as a fuzzy set over a two-dimensional domain, similar to how a fuzzy real number is a fuzzy set over the real domain. Consequently, all points are assigned membership grades (values in the range [0,1]) indicating the extent to which each point belongs to the region. The higher a membership grade is for a point, the more a point belongs to the region; a membership grade 1 indicates it belongs fully to the region, a membership grade 0 indicates that it does not belong to the region. Points with membership grade 0 are not part of the fuzzy set, as is the case in fuzzy set theory.

2.2. DEFINITIONS

The membership grade assigned to each location indicates the extent to which this location belongs to the fuzzy region (the fuzzy set is interpreted veristic: all points belong to the region, but some to a greater extent than others).

$$\tilde{A} = \{(p, \mu_{\tilde{A}}(p))\}$$

where

$$\mu_{\tilde{A}} : U \to [0,1]$$
$$p \mapsto \mu_{\tilde{A}}(p)$$

Here, U is the universe of all locations p; the membership grade $\mu_{\tilde{A}}(p)$ expresses the extent to which p belongs to the region. An illustration of a fuzzy region $\tilde{A}$ can be seen on Figure 3; the shade of grey is related to the membership grade: a membership grade 1 is indicated by black, 0 by white, and shades in between indicate values in between: the higher the grade, the darker the colour.

This approach differs from the egg-yolk and broad boundary approaches mentioned in that it provides additional information regarding the boundary points. Contrary to the previous methods, there now is a distinction possible between the points p_1 and p_2: p_1 is assigned a higher membership grade than p_2, which means that p_1 belongs more to the region than p_2 does. A consequence of our definition is that the boundary of a region is not part of the definition. When it is required – for instance for topology – it can be derived from membership grades of the elements of the fuzzy region. This will be explained further on.

2.3. OPERATIONS

For fuzzy regions as defined above, a number of operations have already been considered. These include operations to combine different fuzzy regions (intersection, union); specific operators from the fuzzy realm (alpha cut); numeric operations (surfacearea, distance to a fuzzy region) and specific geographic operations (minimum bounding rectangle, convex hull). Other operations are under development (i.e., centre of gravity, buffer). By means of an example, just a few of these operations will be given.

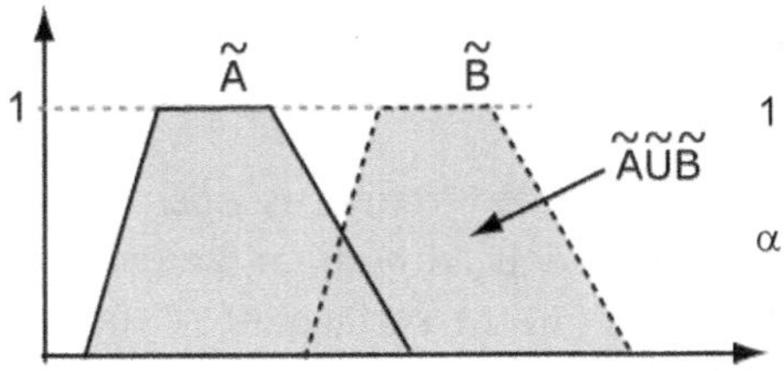

Figure 4. The union of two fuzzy regions; the union is represented by the shaded area

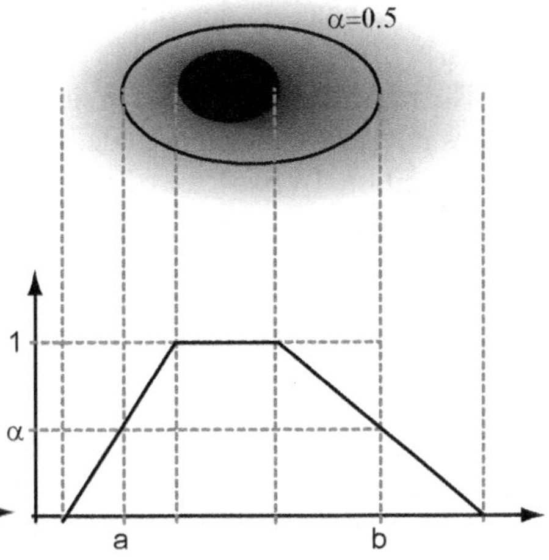

Figure 5. Illustration of the α-cut at 0.5

2.3.1. *Union of fuzzy regions*

Consider two fuzzy regions $\tilde{A}$ and $\tilde{B}$. The union of two fuzzy sets is commonly performed by means of a T-norm, which is also the case for fuzzy regions. A T-norm is a function with two arguments, i.e., commutative, monotone and associative. It also has a null element and an identity element (Klir and Yuan, 1995).

The union of the fuzzy regions $\tilde{A}$ and $\tilde{B}$ is defined using a T-norm and will yield a new fuzzy region.

$$\tilde{A} \tilde{\cap} \tilde{B} = \left\{\left(p, \mu_{\tilde{A}\tilde{\cap}\tilde{B}}(p)\right) : \mu_{\tilde{A}\tilde{\cap}\tilde{B}}(p) = T\left(\mu_{\tilde{A}}(p), \mu_{\tilde{B}}(p)\right)\right\}$$

This is illustrated on Figure 4, where the cross sections of two fuzzy regions $\tilde{A}$ (solid line) and $\tilde{B}$ (dashed line) are given. The cross section of their union is given by the shaded area. The intersection is completely analogue.

2.3.2. *α-cut of a fuzzy region*

In fuzzy set theory, the α-cut serves many purposes, both in defining operations but also in defuzzifying the set (Klir and Yuan, 1995). As a fuzzy region is a fuzzy set over a two-dimensional domain, the definition of the α-cut is straight forward.

$$\tilde{A}_{\tilde{\alpha}} = \left\{(p,1) : \mu_{\tilde{A}}(p) > \alpha, p \in U\right\}$$

As an example, the strong α-cut is given, derived from it is the support (i.e., the strong α-cut at level 0); but the weak α-cut (in which the condition is

$\mu_{\tilde{A}}(p) \geq \alpha$) and its derived respectively kernel (weak α-cut at level 1) are similar.

The result is a fuzzy region, in which all membership grades equal 1, but it does not pose a problem to deduce the crisp boundary and treat this region as a traditional crisp one. This is illustrated on Figure 5, where a fuzzy region is drawn, along with the line where its membership grade equals 0.5. The result of the α-cut will be the crisp region with this same outline.

2.3.3. *Surface calculation*

The meaning of the fuzzy surface area of a fuzzy region is open to two different interpretations. These interpretations depend both on the origin of the fuzziness, but also what meaning is given to the surface area. The first interpretation is the extension of the area for fuzzy regions. The second interpretation considers the surface area to be a sort of cardinality, and matches the definition of fuzzy cardinality.

2.3.3.a. Interpretation 1

The fuzzy surface area $\tilde{S}$ in the first interpretation will result in a fuzzy number that represents the possible surface areas. To obtain the fuzzy number, first, all possible surface areas for the given region must be considered; this is done using the α-cut. With each possible surface area, an appropriate membership grade is associated.

$$\tilde{S}_1(\tilde{A}) = \left\{\left(S(\tilde{A}_\alpha), \alpha\right), \forall \alpha\right\}$$

This interpretation is useful for determining the surface area of a fuzzy region, within a complete fuzzy system (a system that can work with fuzzy numbers). An illustration can be seen on Figure 6: consider a fuzzy region consisting of two squares of size a^2; in one of them all points have membership grade 0.5, in the other all points have membership grade 1. This interpretation yields a fuzzy number, which is 1 for the area a^2 (the area of the α-cut at level 1), and 0.5 for the area $2a^2$ (the area of the α-cut at level 0.5).

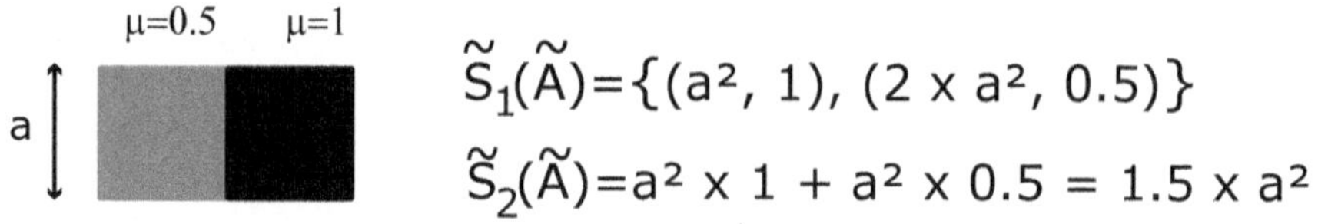

Figure 6. Illustration of the fuzzy surface calculation in both interpretations

2.3.3.b. Interpretation 2

The second interpretation for a fuzzy surface area matches the definition of fuzzy cardinality (Klir and Yuan, 1995) completely. It yields a crisp number, which is a representation for the number of elements. This is done by considering all the points, where the membership grade for each point determines how much it will contribute: a point with a membership grade 0.5 will only contribute half of what a point with membership grade 1 will contribute)

$$\tilde{S}_2(\tilde{A}) = \int_{p \in U} \left(p \times \mu_{\tilde{A}}(p)\right)$$

This interpretation can be used when the results need to be processed by a non-fuzzy system, or when the fuzzy calculations would take too long. While it implies a loss of information, this fuzzy surface area includes some information from the fuzzy locations.

The calculation is also illustrated on Figure 6. The square with membership grade 0.5 only counts for half the amount the square with membership grade 1 does, even though they are the same size.

2.3.4. *Minimum bounding rectangle*

For crisp regions, the minimum bounding rectangle is the smallest rectangular region (with sides parallel to the reference axes) that contains the original region (Rigaux et al., 2002). Consequently, the minimum bounding rectangle of a fuzzy region might be a difficult concept. If the region is not precisely known, how can its bounding rectangle be? Because of this, we introduce the concept of a fuzzy bounding rectangle. Simply put, the fuzzy bounding rectangle is a fuzzy region, so that its α-levels are bounding rectangles for the corresponding α-levels of the region it was constructed for. Consider $MBR(A)$ the notation of the crisp bounding rectangle of a crisp region A, and $M\tilde{B}R(\tilde{A})$. The above property translates to:

$$M\tilde{B}R_\alpha(\tilde{A}) = MBR(\tilde{A}_\alpha), \forall \alpha$$

Figure 7. Illustration of the fuzzy bounding rectangle for fuzzy regions

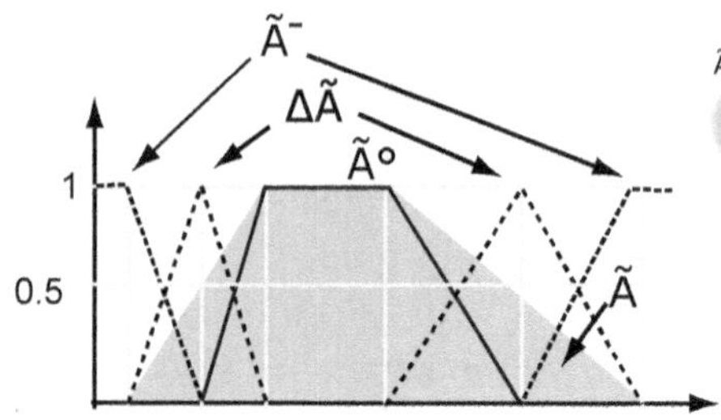

Figure 8. Cross section of the topology concepts

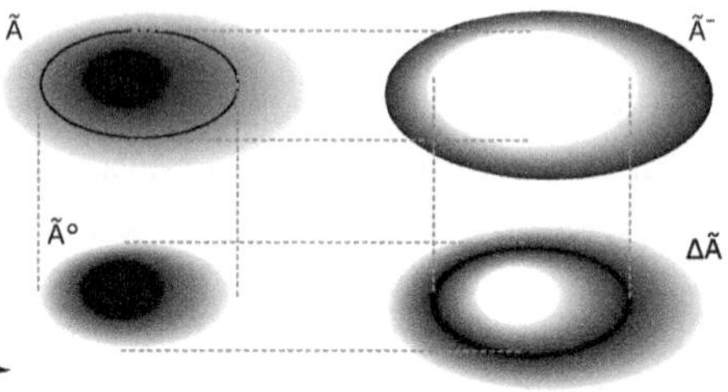

Figure 9. Graphical illustration of the interior, boundary and exterior of a fuzzy region

Basically, the bounding rectangle of each possible α-level, which is a crisp region, is considered. The fuzzy bounding rectangle then is constructed from all of these:

$$\widetilde{MBR}(\widetilde{A}) = \{(p, \mu_{\widetilde{MBR}(\widetilde{A})}(p))\}$$

Where

$$\mu_{\widetilde{MBR}(\widetilde{A})} : U \to]0,1]$$
$$p \mapsto \sup\{\alpha : p \in \widetilde{MBR}(\widetilde{A}_\alpha), \forall \alpha \in]0,1]\}$$

This is illustrated on Figure 7, where on the left side a fuzzy region is shown, and on the right side its fuzzy bounding rectangle is drawn.

2.3.5. *Convex hull*

The convex hull is defined similar to the bounding rectangle. The notion of a fuzzy convex hull is introduced, which is a fuzzy region for which every α-cut represents the convex hull of the corresponding α-cut of the original region. The fuzzy convex hull is defined analogous to the fuzzy MBR.

2.3.6. *Topology*

In order to model the topology for fuzzy regions, some additional concepts are required. Traditionally, topology makes use of the interior, exterior and boundary of a region, and uses intersections of these concepts to describe the topology for the region. To define topology for fuzzy regions, the fuzzy equivalents of these concepts need to be defined first.

The boundary of a fuzzy region $\widetilde{A}$ will be defined as:

$$\Delta\widetilde{A} = \bigcup_{\alpha \in]0,1]} \{(p, 2(0.5 - |0.5 - \alpha|)) : p \in \partial\widetilde{A}_\alpha\}$$

Here, ∂A is the notation for the boundary of the crisp region A (bear in mind that $\tilde{A}_\alpha$ is a crisp region).

Then interior of a fuzzy region $\tilde{A}$ will be defined as:

$$\tilde{A}^\circ = \{(p, \mu_{\tilde{A}^\circ}(p))\}$$

Where

$$\begin{aligned} \mu_{\tilde{A}^\circ} : U &\rightarrow [0,1] \\ p &\mapsto \begin{cases} 0 & \mu_{\tilde{A}}(p) \leq 0.5 \\ 1 - \mu_{\Delta\tilde{A}}(p) & \text{elsewhere} \end{cases} \end{aligned}$$

Then exterior of a fuzzy region $\tilde{A}$ will be defined as:

$$\tilde{A}^- = \{(p, \mu_{\tilde{A}^-}(p))\}$$

Where

$$\begin{aligned} \mu_{\tilde{A}^-} : U &\rightarrow [0,1] \\ p &\mapsto \begin{cases} 0 & \mu_{\tilde{A}}(p) \geq 0.5 \\ 1 - \mu_{\Delta\tilde{A}}(p) & \text{elsewhere} \end{cases} \end{aligned}$$

These concepts – unlike their crisp counterparts – are also fuzzy regions. Completely analogue to the crisp topology, they are used to create an intersection matrix. The intersection matrix is a 3×3 matrix which holds all possible intersections between the boundary, interior and exterior of two regions.

In the crisp nine-intersection model, the matrix elements are considered to be 0 if the intersection is empty and 1 if it is not. In our approach, the matrix elements are deduced from each intersection: each matrix element is the value of the highest membership grade occurring in the intersection. An example of such a matrix element is:

$$\text{height}(\mu_{\tilde{A}\tilde{\cap}\tilde{B}})$$

Where *height* of a fuzzy set X is defined (Kerre, 1991) as:

$$\text{height}(X) = \sup_p \left(\mu_X(p)\right)$$

Note that matrix elements are no longer limited to {0,1}, but can have any value in the range [0,1]. This in turn will impact how the intersection matrices ought to be interpreted.

Using the above definitions of a fuzzy region (consider regions $\tilde{A}$ and $\tilde{B}$), the above definitions for interior, exterior and boundary and using the union of fuzzy regions, the nine-intersection matrix becomes:

$$\begin{pmatrix} h(\tilde{A}^{\circ} \tilde{\cap} \tilde{B}^{\circ}) & h(\tilde{A}^{\circ} \tilde{\cap} \Delta\tilde{B}) & h(\tilde{A}^{\circ} \tilde{\cap} \tilde{B}^{-}) \\ h(\Delta\tilde{A} \tilde{\cap} \tilde{B}^{\circ}) & h(\Delta\tilde{A} \tilde{\cap} \Delta\tilde{B}) & h(\Delta\tilde{A} \tilde{\cap} \tilde{B}^{-}) \\ h(\tilde{A}^{-} \tilde{\cap} \tilde{B}^{\circ}) & h(\tilde{A}^{-} \tilde{\cap} \Delta\tilde{B}) & h(\tilde{A}^{-} \tilde{\cap} \tilde{B}^{-}) \end{pmatrix}$$

Where $h(X)$ is a shorthand notation for the height(X) of a fuzzy set X. A major difference between this intersection matrix and the intersection matrices for both crisp regions as for regions with undetermined boundaries is that in the above matrix, the elements are no longer limited to {0,1}, but are in the range [0,1]. For a case study of the topology, we refer to Verstraete, 2006c.

3. Implementable Models

While the fuzzy region model is interesting as a concept, it cannot be implemented as such. There can be an infinite number of locations, and associating a membership grade with each of them is not attainable in practice. Not only the structure, but also the operations suffer from this difficulty. As illustrated above, some operations require information from each element of the set, which is also not desirable in practice. As illustrated above, many operations for fuzzy regions are relatively straightforward. However, as straightforward as the theory is, it leaves a lot to be desired when it comes to implementability: in theory it is not a problem to define regions and operations using each element of a possibly infinite set; in practice, such definitions are not useful. In order to make the concept of fuzzy regions practically feasible, two implementation models have also been developed. The first is based on the concept of bitmaps in GIS, the second utilises triangulated networks.

3.1. BITMAP MODELS

3.1.1. *Concept*

The use of bitmaps in GIS systems to model data spread over a geographic region is known practice (Shekhar and Chawla, 2003). A bitmap partitions the region in a limited, finite number of cells. With each of the cells, a value that is said to be representative for all the locations in this cell is associated. For our purpose, the associated value will be a membership grade.

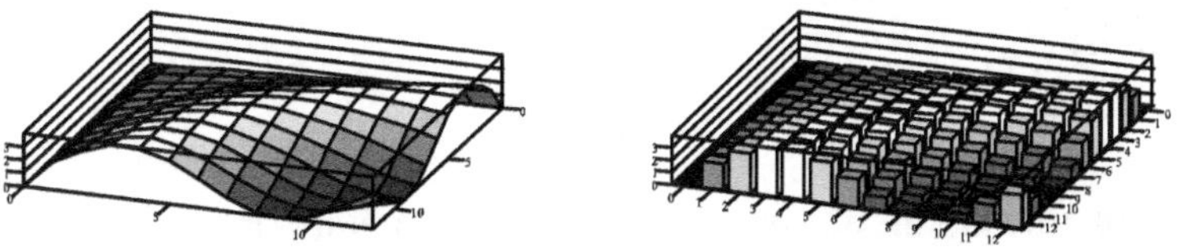

Figure 10. Illustration of the fuzzy bitmap concept: the continuous membership grades are on the left, a bitmap approximation on the right

Conceptually, the bitmap approach is a way of partitioning the infinite number of points into a finite number of sets. As there in a finite number of elements, the operations as defined in the theory require very little modification to be applicable on bitmaps.

In the examples, only bitmaps with rectangular cells are considered. There is no requirement regarding the size of the cells, so cells of different sizes can occur in the bitmap. In some GIS systems, bitmaps with hexagonal cells are used; it should not be a problem to apply the presented techniques to allow such hexagonal cells to be used as well.

3.1.2. *Definition*

In order to define a bitmap, first some required definitions will be provided. A subset $c \subseteq U$ (where U is the universe of all points) is called a cell if it is convex, i.e.,

$$\forall p_1, p_2 \in c, \exists p_3 \in c : \frac{\vec{p}_1 + \vec{p}_2}{2} = \vec{p}_3$$

The cell is the smallest unit known to the bitmap; for some operators the centre point of a cell is used as the reference point for this cell. This point will be denoted as p_c.

A grid – in this context – partitions the region of interest R in a finite collection of disjoint cells:

$$G = \left\{ (c \subseteq U) : \forall c_1, c_2 \in G : c_1 \cap c_2 = \varnothing ; \bigcup_{c_i \in G} c_i = R \right\}$$

Basically, the grid determines the relative position of the various cells that make up the bitmap. As a last step, the membership grade needs to be associated with the cells, which yields:

$$\begin{aligned} \mu_{\tilde{B}} : G &\to [0,1] \\ c &\mapsto \mu_{\tilde{B}}(c) \end{aligned}$$

The definition of a fuzzy bitmap B using grid G and membership function $\mu_{\tilde{B}}$ then is:

$$\tilde{B} = \{(c_i, \mu_{\tilde{B}}(c_i)) : c_i \in G\}$$

The implementation of the fuzzy bitmap structure in a GIS system is relatively easy, about the only thing necessary is to allow the associated values to be limited to the range [0,1]. Of course, the operators necessary to provide functionality for fuzzy bitmaps should still be added.

For a detailed overview of the operations on fuzzy regions in a bitmap representation, we refer to Verstraete et al. (2005) and Verstraete et al. (2006a).

3.2. TRIANGULATED IRREGULAR NETWORKS

3.2.1. *Concept*

Triangulated irregular networks are a commonly used structure in GIS systems (Shekhar and Chawla, 2003). Traditionally, they are used to model features spread over a larger region; the best example is field elevation. Unlike bitmaps however, the triangulated irregular networks are continuous structures. A limited number of points (called *datapoints*) will have an associated value; using these points a *Delaunay* triangulation is constructed. This yields a number of triangles and edges: for each point (that is not a datapoint) an associated data value is calculated using the data of the three corner points of the triangle containing tee point and normal linear interpolation.

This structure has also been adopted for use as a fuzzy region structure. The associated datavalues are membership grades, just like the membership grades that were assigned to the cells of the bitmap. The interpolation will yield membership grades for all points. For some operations it is required to specify edges that must be in the result, for this a *constrained Delaunay*

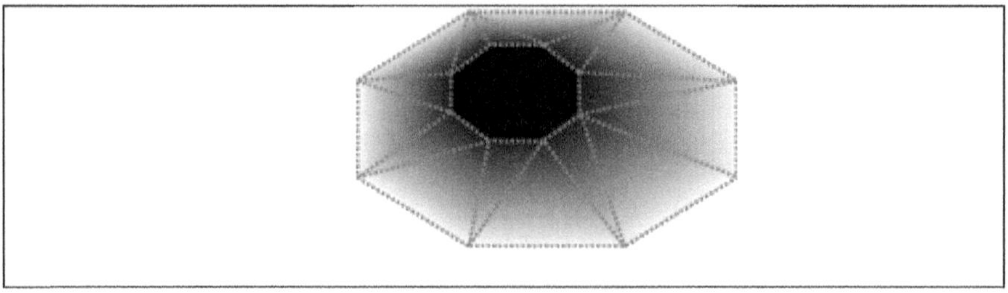

Figure 11. Representation of a fuzzy region using a fuzzy TIN

triangulation is used. Due to the fact that the constrained Delaunay can enforce edges to be in the triangulation, a constrained Delaunay triangulation is no longer a Delaunay triangulation. Both triangulations will be referred to as TIN for the remainder of the paper.

3.2.2. *Definition*

A fuzzy TIN can be defined using a set of datapoints, required edges (in case of a constrained Delaunay) and a mapping function. From these points and edges, a unique TIN can always be constructed. For notational purposes, the fuzzy TIN will be characterised by a set of data points, edges (all edges in the TIN) and triangles (all triangles that occur in the TIN).

$$TIN = [(P, E, T) f]$$

Where P is a set of datapoints on which the TIN is constructed, E is a set of edges (including both the edges obtained through a Delaunay triangulation, as the edges required to be in the result in the case of a constrained Delaunay triangulation), and T is a set of triangles that make up the TIN. The function f is a mapping function defined as:

$$\begin{aligned} f : P &\to [0,1] \\ p(x,y) &\mapsto f(p(x,y)) \end{aligned}$$

Based on the linear interpolation as is applied on a TIN and the mapping function f, the membership function for a fuzzy region $\tilde{A}$ can be defined as

$$\begin{aligned} \mu_{\tilde{A}} : U &\to [0,1] \\ p(x,y) &\mapsto \begin{cases} f(p(x,y)) & \text{if } p(x,y) \in P \\ -\frac{A}{C}x - \frac{B}{C}y - \frac{D}{C} & \text{if } p(x,y) \in R \setminus P \\ 0 & \text{if } p(x,y) \notin R \end{cases} \end{aligned}$$

where R represents the region of interest of the TIN – the region of interest is similar in interpretation to the outline of a crisp region (the polygon) and is immediately deduced from the TIN: the triangulation will automatically yield an outline for any given set of points and edges – and A, B, C and D are the parameters of the equation $Ax + By + Cz + D = 0$ of the plane containing the three points $p_1(x_1, y_1, z_1)$, $p_2(x_2, y_2, z_2)$ and $p_3(x_3, y_3, z_3)$ (with the understanding that $z_j = f(p_j(x_j, y_j))$, j = 1,2,3). Of course, the triangle $p_1(x_1, y_1, 0)$, $p_2(x_2, y_2, 0)$ and $p_3(x_3, y_3, 0)$ is a triangle of the TIN and $p(x,y,0)$ is inside or on an edge of this triangle.

$$
\begin{aligned}
A &= y_1(z_2 - z_3) + y_2(z_3 - z_1) + y_3(z_1 - z_2) \\
B &= z_1(x_2 - x_3) + z_2(x_3 - x_1) + z_3(x_1 - x_2) \\
C &= x_1(y_2 - y_3) + x_2(y_3 - y_1) + x_3(y_1 - y_2) \\
D &= -Ax_1 - By_1 - Cz_1
\end{aligned}
$$

The points $p_1(x_1, y_1, 0)$, $p_2(x_2, y_2, 0)$ and $p_3(x_3, y_3, 0)$ in the *XY*-plane should not be colinear, which is guaranteed by the fact that no Delaunay triangulation (or even in a constrained Delaunay triangulation) would result in a triangulation containing such a degenerate case. For the remainder of the paper, TIN will be used to refer to both a TIN obtained through a Delaunay or through a constrained Delaunay triangulation.

For a detailed overview of the operations on fuzzy regions in a TIN representation, we refer to Verstraete et al. (2005) and Verstraete et al. (2006b).

4. Applications

The fuzzy structures as defined, be it represented as a bitmap or as a TIN, can have a multitude of applications. Depending on the application, the interpretation of the associated membership degrees can also differ.

4.1. FUZZY REGIONS

A first application is to use the fuzzy region to model region a region with an imprecise boundary. Each point of the fuzzy region has membership grade, and grades are interpreted veristic. This means that all points with a grade strictly greater than 0 belong to the region; the grade represents the extent to which the point belongs to the region. The requirement to model a region to which points can partly belong, can have a number of applications: current data could only be acquired as an estimate, the data could concern predictions about future data and sometimes could be a result of tracing back in the past (especially, if there is no written history of the changes to reach the current situation). A number of possible sources of fuzziness will now be considered.

4.1.1. *Inherent fuzziness*

Some data that are modelled in geographic systems are inherently fuzzy, even though they are currently modelled crisply. The fuzziness in this case does not stem from limited measurements, but is present in the real world. The composition of the soil is such an example: where does one type of soil (e.g., clay) stops, and another type (e.g., sand) begins? Along the same lines, the spread of a contaminant (for instance oil pollution) in the ground is inherently fuzzy.

4.1.2. *Estimated data*

It happens that the real data is crisply defined, but it is virtually impossible to collect. Consider the temperature: at a given time, every location will have an exact temperature, but of course it is impossible to measure this temperature at every location. Similarly, the territory of various animals in the wild, while quite commonly crisply defined by the animals themselves has to be estimated by people studying the animals.

4.1.3. *Predictions/backtracking*

Obviously, the future is not certain. Any data that involves predictions about future situations is therefore prone to uncertainty or imprecision. Similarly, if data from the past are modelled, chances are there is no accurate record of the actual situations. Modelling past data can therefore also benefit from incorporating the fuzziness.

4.2. FUZZY POINTS

Fuzzy regions can also be used to represent points at imprecisely known locations. In this sense, the fuzzy regions model possible locations for the fuzzy point being modelled. Consequently, the membership grade for each location represents the extent to which this location is a possibility for the fuzzy point. The major difference with the above interpretation is that now the interpretation is possibilistic: only one location is valid, but it is unknown at this point which location it is (this interpretation bears much resemblance to the concept of fuzzy numbers). Because of this change in interpretation, some operators might have to change: the surface area of the point itself is 0, but one can say that the possible locations are spread over a given surface area; the distance to a fuzzy point also differs from the distance to a fuzzy region.

4.3. FUZZY DEGREES OF TRUTH

Even in crisp databases, it can be useful to allow fuzzy queries. In this situation, all the data in the database is crisp (so no changes to the database are required), but the query engine allows for fuzzy concepts (e.g., a distance notion of *closeby*). The third interpretation for fuzzy sets – as mentioned in Dubois – is as degrees of truth. The result of a fuzzy query on crisp data can then a number of locations; consider the query *all the locations close to* a given location, each with a degree of truth. This degree can be represented by an associated membership grade, and thus in the form of a fuzzy region.

5. Conclusion

In this paper, an overview of our work in the field of fuzzy spatial data types was given. The work includes a full theoretical basis, on which two different implementable models have been based. For both of these models, a number of operations have been defined and tested using various developed prototype implementations. They yield an intuitive result, consistent with the theoretical model while still appearing feasible to implement. Further testing is needed regarding the performance of complex fuzzy structures; for this purpose, a full implementation of both models is in the works. Other means of visualising fuzzy regions are also still under investigation.

References

Clementini E., Di Felice P., 1994, An Algebraic Model for Spatial Objects with Undetermined Boundaries, GISDATA Specialist Meeting – revised version.

Cohn A. G., Gotts N. M., 1994, Spatial Regions with Undetermined Boundaries, Procedings of the 2nd ACM Workshop on Advances in GIS, 52–59.

Dubois D., Prade H., 1997, The Three Semantics of Fuzzy Sets, Fuzzy Sets and Systems, 90: 141–150.

Hallez A., Verstraete J., De Tré G., De Caluwe R., 2002, Contourline Based Modeling of Vague Regions; Proceedings of the 9th International Conference on Information Processing and Management of Uncertainty in Knowledge-Based Systems (IPMU) 2002, July 1–5, Annecy, France.

Kerre E. E., 1991, Introduction to the basic principles of fuzzy set theory and some of its applications; Etienne E. Kerre, Communication and Cognition.

Klir G. J., Yuan B., 1995, Fuzzy sets and fuzzy logic: Theory and applications, New Jersey: Prentice-Hall.

Morris A., 2001, Why Spatial Databases Need Fuzziness, Proceedings of Nafips 2001 2446–2451.

Rigaux P., Scholl M., Voisard A., 2002, Spatial Databases with Applications to GIS, Morgan Kaufman Publishers.

Shekhar S., Chawla S., 2003, Spatial Databases: A Tour, Pearson Education.

Vertraete J., Van Der Cruyssen B., De Caluwe R., 2000, Assigning Membership Degrees to Points of Fuzzy Boundaries, Proceedings of the 19th International Conference of the North American Fuzzy Information Processing Society – Nafips, 444–447.

Verstraete J, De Tré G., De Caluwe R., Hallez A., 2005, Field Based Methods for the modelling of Fuzzy Spatial Data, Fuzzy modeling with Spatial Information for Geographic Problems, F. Petry., V. Robinson, M. Cobb, editors, Springer, 41–69.

Verstraete J, Hallez A., De Tré G., 2006a, Bitmap Based Structures for the modeling of Fuzzy Entities, Control & Cybernetics: Geographic Information Systems and Decision Support: New Approaches and Applications, J. Horak, J. Kacprzyk, A. Morris, J. W. Owsinski and S. Zadrozny, guest editors; to appear.

Verstraete J., De Tré G. Hallez A., De Caluwe R., 2006b, Using TIN-based structures for the modelling of fuzzy gis objects in a database, IJUFKS (special issue on "Intelligent Fuzzy Information Systems: Beyond the Relational Data Model"), to appear.

Verstraete J., Hallez A., De Tré G. Matthé T., 2006c, Topological relations on fuzzy regions: intersection matrices, Proceedings of the Information Processing and Management of Uncertainty conference (IMPU), 2104–2111.

MAPPING THE ECOTONE WITH FUZZY SETS

CHARLES ARNOT*
Department of Geography, University of Leicester, Leicester, LE1 7RH, United Kingdom

PETER FISHER*
Department of Information Science, City University, Northampton Square, London EC1V 0HB, United Kingdom

Abstract. Ecotones are the zones of transition between patches of different ecological character. In landscapes where humans manage the land, they can be sharp or abrupt in space, but in semi-natural environments they more typically occupy space showing an intergrade between one ecological area and another. Semi-natural ecotones are, therefore, poorly treated by the traditional Boolean mapping of vegetation and land cover with sharp spatial boundaries which are implicit in it. Fuzzy sets provide a means by which ecotones can be represented as 2-dimensional spatial objects; fuzzy type 2 sets provide a further dimension to this characterization which is more in keeping with the higher order fuzzy nature of ecotones. This paper presents a methodology for representing ecotones as fuzzy objects with examples of a forest-savanna ecotone from Bolivia.

Keywords: ecotone, forest, fuzzy change analysis, fuzzy sets, fuzzy logic, land cover mapping, savanna, type 2 fuzzy sets

1. Introduction

Land cover can be interpreted as a series of simple structures principally composed of patches outlined by boundaries through which other structures, such as corridors can be distinguished (Forman and Godron, 1986; Forman, 1995; Cadenasso et al., 2003b). Historically these patches have been conceived and mapped as Boolean areas of perfect internal homogeneity. This is in contrast to the theoretical properties of patches which describe an

* To whom correspondence can be addressed: email p.fisher@city.ac.uk, charlie.arnot@gmail.com

A. Morris and S. Kokhan (eds.), Geographic Uncertainty in Environmental Security, 19–32.

object defined by context in terms of relative scale and relative thematic homogeneity. In other words, a patch consists of an amalgamation of other patches at different *spatial* and *thematic* scales. A patch of woodland can be broken down into smaller patches, that is, by woodland type (thematic resolution), or may appear differently, that is, as a number of small woodland islands close together, when viewed at a more magnified spatial resolution. In the homogeneous patch model, therefore, the apparent homogeneity of patches masks the underlying variation occurring within them. It is this variation that is of interest because it contains more information about the landscape than the homogeneous Boolean model. It has been argued elsewhere in this volume (Bastin et al. and Fisher and Arnot) that vegetation communities and land covers are suitable for modelling by fuzzy sets due both to identification and spatial fuzziness. In this article we assume that the fuzzy set model is accepted. The issue addressed is therefore how the zone of transition between the location of one community and another which is equivalent to the boundary is modelled by fuzzy sets. In ecology this transition zone is widely referred to as the ecotone. More correctly a zone of transitions is called an ecocline (Kent et al., 1997), but the more widely used term the ecotone is usually used here.

Kent et al. (1997) point out that in all the research in landscape ecology and vegetation mapping very little seems to have been done on explicit modelling of the ecocline as a spatial entity. Most work on the ecotone has been to identify the optimal position of the abrupt Boolean ecotone within a zone of an ecocline, and novel methods such as wombling have been introduced to achieve this (Fortin and Edwards, 2001; Fortin et al., 2000).

Research on the ecotone is discussed in section 2. In section 3 its representation in fuzzy sets is outlined (both type 1 and type 2 fuzzy sets), and in section 4 the study site and methods used in analysis are introduced. Section 5 contains some results of this modelling for an area of Bolivia.

2. Ecotones

Ecologists, and Landscape ecologists in particular, seek to understand how landscape pattern and processes are linked (Watt, 1947; Legendre and Fortin, 1989; Levin, 1992; Grimm, et al., 1996; Wiegand et al., 2003; Turner, 1989; Turner, 2005). Ecotones are important landscape structures from this perspective because they are associated with numerous ecological factors i.e., edge effects, corridor effects, barrier effects, that are dependent on the ecotones spatial characteristics (Hansen, et al., 1988; Fagen et al., 1999; Lidicker, 1999; Fortin et al., 2000; Ries et al., 2004). As a consequence there is abundant research into ecotones, edge effects and associated processes, however this research is usually characterised by a lack of spatial representation of the ecotone itself, and indeed, there is some confusion as to the definition of ecotones (Strayer et al., 2003). Ecologists tend to examine processes occurring either within transition zones, or within areas of homogeneity, seldom linking the two (Cadenasso et al., 2003b).

Landscape ecologists impose a structure of patches and patch boundaries (ecotones) in terms of land cover. Traditionally, landcover ecotones have been represented using a Boolean model. Within this framework ecotones are conceptualised as structures with no area, having only a spatial dimension of length (Fortin et al., 2000). Although the spatial form of ecotones can be abrupt, that is, a field margin, it may also be graduated (Kent et al., 1997; Lidicker, 1999; Bowersox and Brown, 2001; Kent et al., 2006). As such ecotones are ideally modelled using a fuzzy rather than Boolean approach.

Currently there is much research into ecotones and boundary processes but this is mainly focused on ecological processes and species distributions rather than specifically mapping ecotones. In particular Alpine treelines have recently become the subject of much research because they contain population processes and spatial patterns that change along environmental gradients which make them ideal for study (Slayter and Noble, 1992; Camero and Guitairrez, 2002; Grytnes, 2003; Wiegand et al., 2006), although there has also been research into many different types of ecotone and ecotonal theory (Bowersox and Brown, 2001; Olsen, 2001; Cadenasso et al., 2003a; Cadenasso et al., 2003b; Fagen et al., 2003; Parmesan et al., 2005; Kent et al., 2006).

There are a variety of approaches taken in detecting and mapping ecotones. Fortin (Fortin and Drapeau, 1995; Fortin et al., 1996; Fortin et al., 2000) has presented many papers on boundary extraction and processes, in particular the use of wombling, as a statistical technique used to identify optimal boundary positions. In a review of boundary detection, Fortin et al. (2000) describe a variety of approaches based on remote sensing (edge detection, segmentation, moving window analysis), statistical approaches (lattice wombling) and modelling. In particular Fortin recognises the limited research in describing ecotones within a landcover context. Fortin et al. (2000) conclude "the challenge is to use the available data and techniques in ways that identify the heterogeneous zones as entities rather than simply to reduce them to a line between adjacent patches" (p. 462).

Research into the spatial characterisation of ecotones as fuzzy objects is sparse and seldom linked to ecological process or even landcover. Burrough (1996) uses fuzzy objects to identify fuzzy boundaries, and Burrough and McDonald (1998) describe how a confusion index was used to extract class boundaries from a four-class fuzzy classification to create fuzzy boundaries. The confusion index is the ratio of the second largest membership value against the first. When the value was small, that is, a high confusion, there was an indication of a boundary location and width, however there was no indication of what the contributing classes were to the boundary. Cheng and Molenaar discuss fuzzy boundaries as parts of fuzzy objects (Cheng and Molenaar, 2000), Foody and Boyd (1999) discuss transition areas in relation to land cover change without explicitly mapping the boundaries. Wang and Hall (1996), building on the work of Leung (1987) acknowledge the inadequacy of vector boundaries in modelling transition zones and suggest a methodology for representing transition zones derived from a minimum operator.

For a comprehensive review of boundary detection in ecology see Kent et al. (2006).

3. Fuzzy Sets and Ecotones

When he introduced the basic premise of fuzzy set theory, that the membership of a location (or case) is described by a real number in the range between 0 and 1, Zadeh (1965) proposed a formal theory of logic for the combination and manipulation of fuzzy sets. The mathematics of this has been expanded and developed to absorb the relaxation of many of the different mathematical assumptions of Boolean set theory (Leung, 1987; Klir and Yuan, 1995).

At the heart of fuzzy set logic are two very simple operations, the fuzzy union and fuzzy intersection which identify the membership of the occurrence of either sets A or B and both (respectively). Given that the ecotone is composed of that location that to some degree belongs to both of two cover types or vegetation communities, Leung (1987) identified that the intersection operator can be used to identify the boundary or ecotone (Figure 1). Therefore, at a specific location, x, the ecotone is the result of applying the minimising intersection operation (Equation 1).

$$\mu(BoundA \cap B)_{x_x} = \min(\mu(A)_x, \mu(B)_x) \quad (1)$$

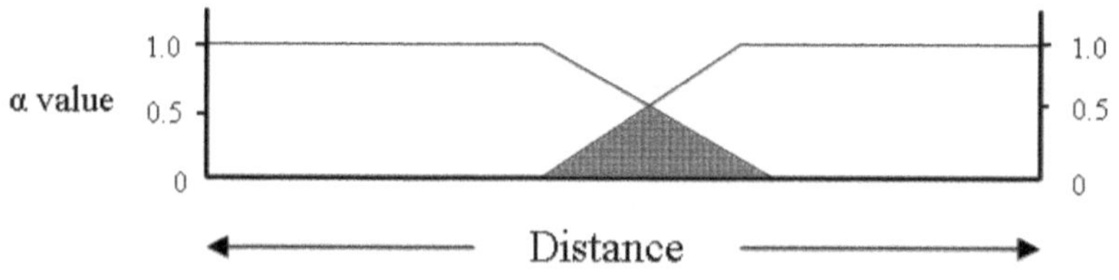

Figure 1. The intersection between two fuzzy sets is indicated by the shaded area, and when the sets are land cover types it is taken to be the intergrade

Working with classified images, the memberships in the two input cover types for which the ecotone is being determined will never be more than 0.5, and so the ecotone set may contain relatively low membership values, but these can be normalised to values in the range 0 to 1 by dividing the intersection value by the superior value of all intersection set (0.5 or some other appropriate and justified value in the range 0 to 1) (Equation 2 or 2A).

$$Norm(\mu(BoundA \cap B)_x) = \frac{\mu(BoundA \cap B)_x}{\sup(\mu(BoundA \cap B))} \tag{2}$$

$$Norm(\mu(BoundA \cap B)_x) = \frac{\mu(BoundA \cap B)_x}{0.5} \tag{2A}$$

The basic set of ecotones will be between pairs of cover types (e.g., A and B), and for any situation with n cover types there will be $\frac{n(n-1)}{2}$ ecotones.

It is possible to map the union of ecotones between all pairs of cover types in the area, using the standard fuzzy set union operator in the area, to yield the direct comparisons with the Boolean ecotones. This is shown in Figure 3.

$$\mu(AllBounds) = \max_{i=1}^{n}(Bound\mu(C_i),) \tag{3}$$

Fisher and Arnot (this volume) have argued that type 2 fuzzy sets (Zadeh, 1975; Mendel and John, 2002) add to the representational power of fuzzy sets by accommodating higher order uncertainty. In the case of ecotones, this may be even more marked. Ecotones should be the locations in the landscape with the maximum amount of doubt in their allocation to a class and it was expected that this will be evident in a type 2 analysis.

4. Study Area and Methods

Following the work of Fisher and Arnot (this volume) we have used a small experimental area of Bolivia to examine the concepts discussed here. We identified four cover types in the area including water, wet and dry savanna and forest, and they were classified with the fuzzy *c*-means classifier (Bezdek, 1981; Bezdek et al., 1984) implemented in Parbat by Lucieer (2004). Following the review of fuzziness values used elsewhere in the literature, and experimentation with the data-set presented here, fuzzy memberships for a type 1 fuzzy set using a fuzziness, m, of 1.6 was used. Following the methods of Fisher et al. (2004) and Fisher and Arnot (this volume), we use the different instances of the type 1 classification to parameterise the type 2 sets, and as in the latter work the fuzziness in the results was varied from 1.3 to 2.5 and the variation in memberships explored and compared.

5. Results

5.1. TYPE 1 ECOTONES

Figure 2 shows maps of two cover types as fuzzy sets (Figure 2A and 2B), the intersection or ecotone (Figure 2C) and the normalised ecotone (Figure 2D). The wet savanna is relatively restricted in the study area while dry savanna is more extensive. The difference between the normalised and unnormalised ecotone is evident in the darkness of the former map

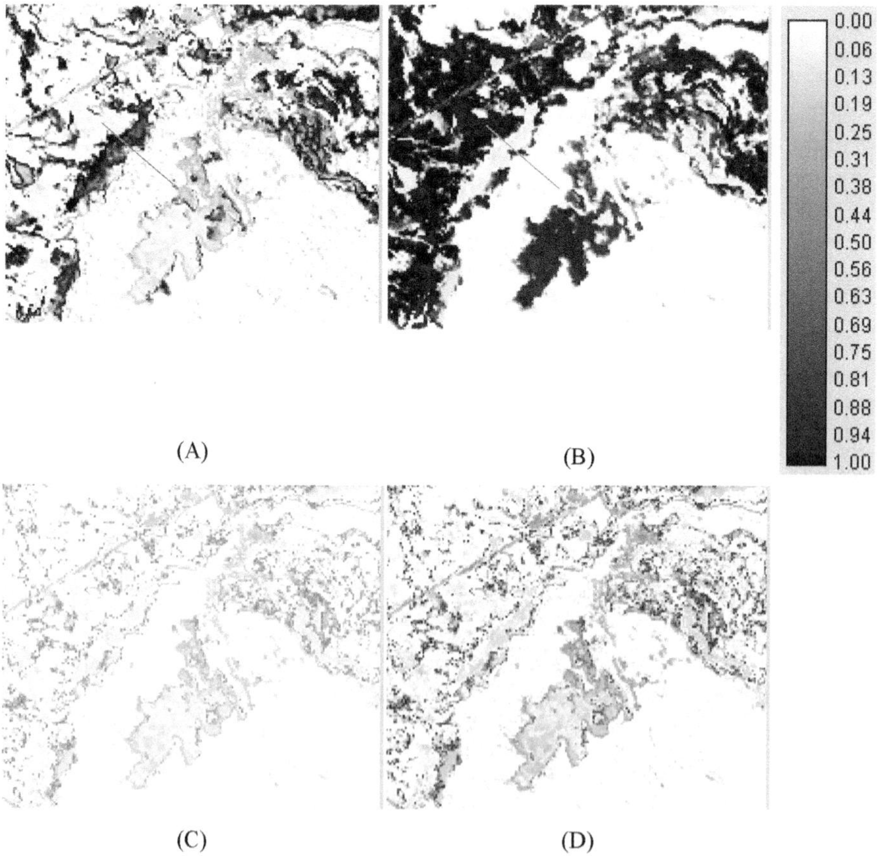

Figure 2. The derivation of a fuzzy ecotone when fuzziness, m = 1.6: (A) the extent of wet savanna; (B) the extent of dry savanna; (C) the fuzzy intersection of the two covers; and (D) the normalised intersection of the two covers

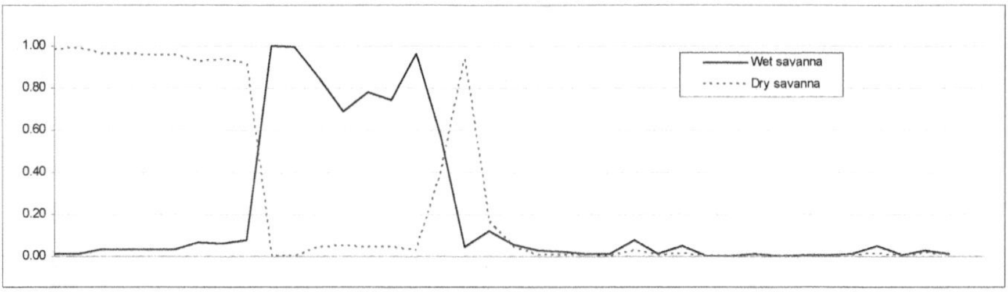

(A) The fuzzy memberships of wet and dry savanna

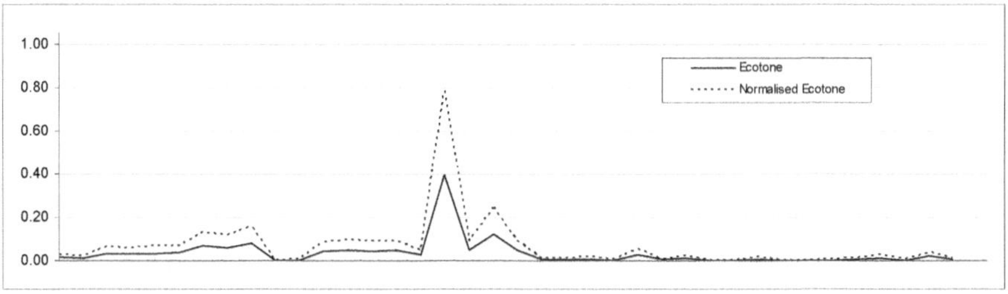

(B) The fuzzy memberships of ecotones between wet and dry savanna

Figure 3. Transects through the fuzzy memberships for the wet and dry savanna types as for the ecotone and normalized ecotone between them

(Figure 2D). It is therefore easier to see patterns in the map and to identify the structure in the ecotone.

In Figure 3 transects are illustrated for the location indicated in Figures 2A and 2B. The cover types are clearly identified with the main area of wet savanna to the left of the transect and another area occurring (actually at one pixel) to the middle. The transition from wet to dry savanna to the left of the area is very abrupt; it actually occurs between one pixel on the transect and the next. On the other hand, the transition to the second occurrence of dry savanna does have a narrow intergrade area; there is one pixel in the intergrade on each side and a single pixel in the dry savanna. The map (Figure 2) suggests that the intergrades illustrated in the transect are very varied, and nowhere could be considered typical.

Figure 4 presents a pair of images which show the fuzzy and the Boolean representations of the ecotone (more correctly the fuzzy ecocline and the Boolean ecotone; Kent et al., 1997). The edge of the image pixels or grid cells is traced out. This is a necessary consequence of the Boolean assumption of traditional mapping; the Boolean ecotone occupies no area, but is a 1D, linear object.

It is apparent that the information about the ecotones is greatly enhanced in the fuzzy representation. Specifically the area occupied by the ecotone (to some degree) is represented and those areas where doubt exists as to the

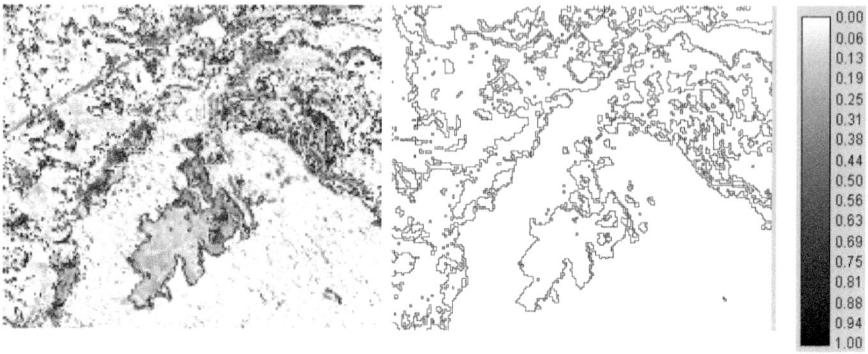

(A) (B)

Figure 4. The full pattern of ecotones in the area when fuzziness, m = 1.6: (A) the union of normalised ecotones for all six pairs of cover types and (B) the equivalent Boolean ecotones

existence of boundaries is represented. The area of dry savanna in the middle of the area (Figure 2B) is shown to have a small residual membership of the class wet savanna (Figure 2A) and so to a small degree to be part of the fuzzy ecotone, especially the normalised ecotone (Figure 2D). On the other hand, the area is ringed by a zone of large membership in the ecotone to all classes (Figure 4A) and has a definite boundary, which is in part crisp, but is much more diffuse to the north and the east. This is partly reflected by the disjoint pattern of a mosaic of boundary in the Boolean representation, but it appears as a smooth transition in the fuzzy representation. Generally, there is no relationship between the apparent mosaic nature of the Boolean boundaries and the wider fuzzy boundaries.

5.2. TYPE 2 FUZZY ECOTONES

An extension of the principle of fuzzy sets is type 2 fuzzy sets which account for higher order vagueness or the uncertainty of the uncertainty. Fisher and Arnot (this volume) discuss the use of type 2 fuzzy sets in the identification of land cover change. Type 2 sets are reviewed by Mendel and John (2002).

Fisher and Arnot (this volume) illustrate and discuss the extent of the type 1 and 2 sets of the basic land cover classes in the study area. They show how the memberships and valuations of the sets are derived and interpreted, especially with respect to reporting of the extent of forest in the study area.

Using the same terminology as Fisher and Arnot (this volume) to interpret the type 2 sets, Figure 5A shows the extent of the core ecotone areas; the minimum fuzzy membership of the union of fuzzy ecotones between paired land cover classes (e.g., Figure 1D) for each of the 13 different valuations of m in the fcm classifier used to derive the type 1

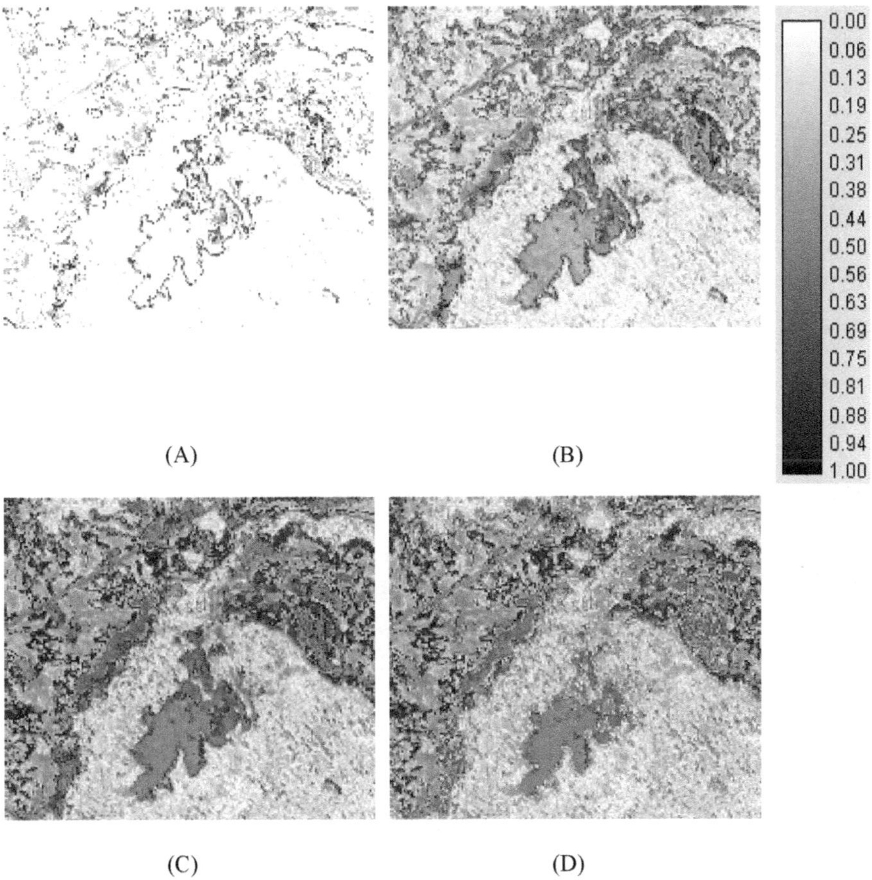

Figure 5. The type 2 fuzzy set of all ecotones: (A) the minimum (core) ecotone; (B) the mean (typical) ecotone; (C) the maximum (generous) ecotone, and; (D) the range of the ecotone values

fuzzy sets. These are the areas which best fit the concept of a zone of intergrade between any land covers. They can be seen to generally be relatively restricted in spatial extent (narrow bands), but occasionally they thicken. They can be seen to be more conservative than the ecotone in the type 1 fuzzy set when the fuzziness, m, is 1.6 (Figure 3A). Specifically the area of dry savanna in the centre of the study area does not have the moderate levels of fuzzy membership in the ecotone – at least one classification does not pick this area out as anything other than dry savanna.

The mean fuzzy membership (the typical ecotone; Figure 5B) shows a much more extensive residual membership in other classes in the core areas of classes than is apparent in the type 1 set when m equals 1.6 (Figure 3A). The generous interpretation of the type 2 set (the maximum membership; Figure 5C) shows that some classes are very poorly defined as pure classes

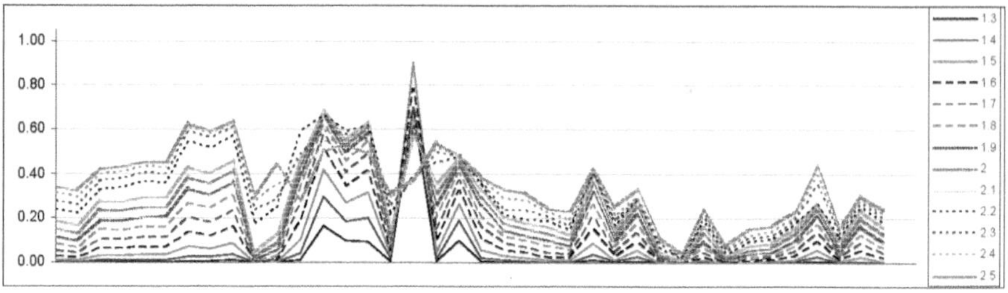

(A) The set of type 1 memberships of the ecotone of all classes

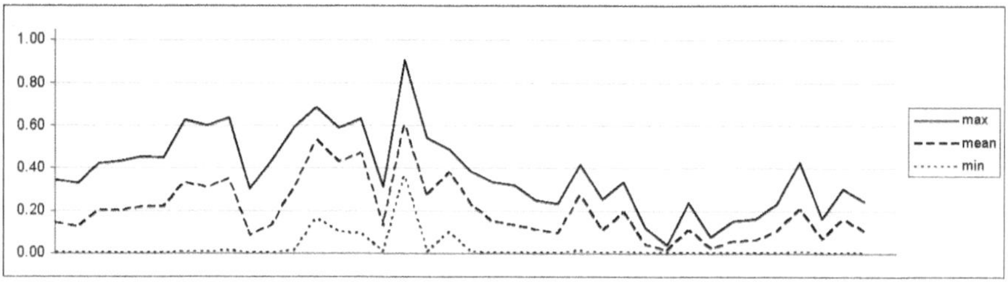

(B) The type 2 set of ecotones of all classes

Figure 6. Transects on the line shown in Figures 1A and 1B and corresponding to those shown in Figure 3

because the membership of at least one ecotone between a pair of classes is towards the middle of the range of possible values. Figure 5D shows the range of memberships in the type 2 sets (simply the difference of the maximum and minimum memberships). This value is very similar to the maximum in most areas because the minimum is 0 in those areas. Where the minimum is larger, the range is less, sometimes much less than the maximum. Note that memberships of all ecotones shown in Figure 5 were derived from normalised ecotones, but are not themselves normalised.

Transects in Figure 6 show that the type 2 ecotone sets are very different from any one type 1 set. In many parts of the study area it is correlated to the value in the 2.5 realisation, but in other parts type 1 sets where m is 1.4, 2.1 and 2.2 are the maximum value. Similarly the type 1 set which contributes the minimum varies along the length of the transect. Furthermore the relationship of the range to the maximum and minimum is emphasised.

The independence of the parameters of the type 2 set from any one type 1 set is further illustrated in the scatter plots in Figure 7. The spread of points in the scatter plots shows some interesting patterns, which suggest that multiple populations are modelled, showing some relationship to the type 1 set on the *x*-axis. However, the type 1 set is never a reliable predictor of the type 2 set values. Indeed, the minimum values are almost entirely

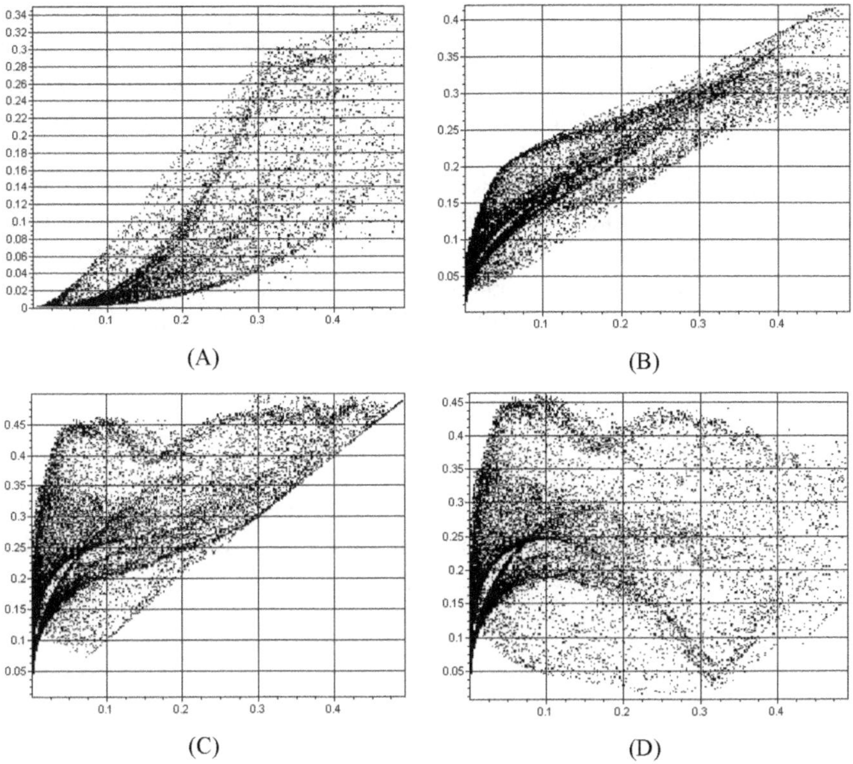

Figure 7. Scatterplots of the fuzzy membership of all ecotones in the classification where fuzziness, m = 1.6, plotted as the x-axis against parameters of type 2 fuzzy sets: (A) the minimum; (B) the mean; (C) the maximum and (D) the range

below the leading diagonal, the mean values spread around it, and the maximum values entirely below it. Most interestingly, the range values spread over the whole area of the scatterplot; not possible value of the range of memberships is excluded by the type 1 membership.

6. Conclusion

Here we have illustrated the ecotone (more correctly ecocline) as a mappable entity in its own right occupying 2D space, and not a traditional 1D phenomenon as it is in the Boolean model of space. We have shown that in a landscape of n classes it is actually possible to generate $\frac{n(n-1)}{2}$ possible ecotones between each pair of classes. From those we derived an aggregate measure of all ecotones by taking the union of all possible ecotones. Furthermore, we have argued that the various mappings of the type 1 sets hold a great deal of information on the variability of the classes along the ecotones between land cover allowing a more detailed interpretation of structure of those ecotones. We have also shown that the parameters of the type 2 set seems to hold still more information and that the variability expressed in this set may be even more interesting as an indication of structure.

Fortin et al. (2000), in a review of boundary detection, explicity call for methodologies in land cover analysis that can define, describe and model boundaries as spatial entities, which is exactly what is presented in this paper. It is however, an experiment in set theoretic modelling of the ecotone. It requires field validation which is beyond both the scope of the current paper.

Acknowledgement

We would like to thank Richard Wadsworth and Jane Wellens for their contributions to preliminary stages of this work. The results and conclusions are entirely the responsibility of the authors.

References

Bezdek, J.C., 1981, *Pattern Recognition with Fuzzy Objective Function Algorithms*, NY: Plenum Press.

Bezdek, J.C., Ehrlich, R., and Full, W., 1984, FCM: The fuzzy *c*-means clustering algorithm, *Computers & Geosciences* **10**: 191–203.

Bowersox, M.A., and Brown, D.G., 2001, Measuring the abruptness of patchy ecotones. A simulation-based comparison of landscape pattern statistics, *Plant Ecology* **15**: 89–103.

Burrough, P.A., 1996, Natural objects with indeterminate boundaries, in Burrough, P.A., and Frank, A.U., eds., *Geographic Objects with Indeterminate Boundaries.* London: Taylor & Francis, pp. 3–28.

Burrough, P.A., and McDonnell, R.A., 1998, *Principles of Geographical Information Systems* 1st ed. Oxford: Oxford University Press.

Cadenasso, M.L., Pickett, S.T.A., Weathers, K.C., Bell, S.S., Benning, T.L., Carreiro, M.M., 2003a, An interdisciplinary and synthetic approach to ecological boundaries *Bioscience* **53**: 717–722.

Cadenasso, M.L., Pickett, S.T A., Weathers, K.C., and Jones, C.G., 2003b. A framework for a theory of ecological boundaries, *Bioscience* **53**: 750–759.

Cheng, T., and Molenaar, M. 2000, The Identification and Monitoring of Objects with Fuzzy Spatial Extent, *International Archives of Photogrammetry and Remote Sensing* **32**: 207–212.

Fagan, W.F., Cantrell, R.S. & Cosner, C. 1999: How Habitat Edges Change Species Interactions. *The American Naturalist*, 2, **153**: pp. 165–182.

Fisher, P.F., Wood, J., and Cheng, T., 2004, Where is Helvellyn? Multiscale morphometry and the mountains of the English Lake District, *Transactions of the Institute of British Geographers* **29**: 106–128.

Foody, G.M., and Boyd, D.S. 1999. Detection of partial land cover change associated with the migration of inner-class transitional zones, *International Journal of Remote Sensing* **20**: 2723–2740.

Forman, R.T.T., 1995, *Land Mosaics; The Ecology of Landscapes and Regions*, Cambridge, University Press.

Forman, R.T.T., and Godron, M., 1986, *Landscape Ecology*, Chichester: Wiley & sons.

Fortin, M.J., and Drapeau, P., 1995, Delineation of ecological boundaries: comparison of approaches and significance tests, *Oiko* **72**: 323–332.

Fortin, M.J., Drapeau, P. and Jacquez, G.M., 1996, Quantification of the spatial co-occurrences of ecological boundaries, *Oiko* **77**: 51–60.

Fortin, M.-J. and Edwards, G. 2001. Delineation and analysis of vegetation boundaries. in: Hunsaker, C.T., Goodchild, M.F., Friedl, M.A., and Case, T.J. eds., *Spatial Uncertainty in Ecology*, NY: Springer, pp. 158–174.

Fortin, M.-J., Olson, R.J., Ferson, S., Iverson, L., Hunsaker, C., Edwards, G., Levine, D., Butera, K., and Klemas, V., 2000, Issues related to the detection of boundaries, *Landscape Ecology* **15**: 453–460.

Grimm, V., Frank, K., Jeltsch, F., Brandl, R., Uchmanski, J. and Wissel, C., 1996, Pattern-oriented modelling in population ecology, *Science of the Total Environment* **183**: 151–166.

Grytnes, J.A., 2003, Species-richness patterns of vascular plants along seven altitudinal transects in Norway, *Ecography* **26**: 291–300.

Hansen, J.A., diCastri, F., and Naiman, R.J., 1988. Ecotones: What and Why? *Biology International* **17** Special Issue.

Kent, M., Gill, W.J., Weaver, R.E., and Armitage, R.P., 1997, Landscape and plant community boundaries in biogeography, *Progress in Physical Geography* **21**: 315–353.

Kent, M., Moyeed, R.A., Reid, C.L., Pakeman, R., and Weaver, R.E., 2006. Geostatistics, spatial rate of change analysis and boundary detection in plant ecology and biogeography, *Progress in Physical Geography* **30**: 201–231.

Klir, G.J., and Yuan, B., 1995, *Fuzzy Sets and Fuzzy Logic: Theory and Applications* Englewood Cliff, NJ: Prentice-Hall.

Legendre, P., and Fortin, M., 1989, Spatial pattern and ecological analysis, *Vegetatio* **80**: 107–138.

Leung, Y.C., 1987, On the imprecision of boundaries, *Geographical Analysis* **19**: 125–151.

Levin, S.A., 1992, The problem of pattern and scale in ecology, *Ecology* **73**: 1943–1967.

Lidicker, W.Z. 1999. Responses of mammals to habitat edges: an overview, *Landscape Ecology* **14**: 333–343.

Lucieer, A., 2004. Parbat: version 0.32. www.parbat.net.

Mendel, J.M., and John R.I., 2002, Type 2 fuzzy sets made simple, *IEEE Transactions on Fuzzy Systems* **10**: 117–127.

Parmesan, C., Gaines, S., Gonzalez, L., Kaufman, D.M., Kingsolver, J., Townsend Peterson, A., 2005, Empirical perspectives on species borders: from traditional biogeography to global change, *Oikos* **108**: 58–75.

Ries, L., Fletcher, R.J., Battin, J., and Sisk, T.D., 2004, Ecological responses to habitat edges: Mechanisms, models, and variability explained, *Annual Review of Ecology Evolution and Systematics* **35**: 491–522.

Slatyer, R.O., and Noble, I.R., 1992, Dynamic of montane treelines, in *Landscape Boundaries: Consequences for Biotic Diversity and Ecological Flows*, A.J. Hansen and F. di Castri, eds., NY: Springer, pp. 346–359.

Strayer, D.L., Power, M.E., Fagan, W.F., Pickett, S.T.A., and Belnap, J., 2003, A classification of ecological boundaries, *Bioscience* **53**: 723–727.

Turner, M.G., 1989, Landscape ecology: the effect of pattern on process. *Annual Review of Ecology and Systematics* **20**: 171–197.

Turner, M.G., 2005, Landscape ecology: What is the state of the science? *Annual Review of Ecology and Systematics* **36**: 319–344.

Wang, F., and Hall, B.G., 1996, Fuzzy representation of geographical boundaries in GIS, *International Journal of Geographical Information Systems* **10**: 573–590.

Watt, A.S., 1947, Pattern and process in the plant community, *Journal of Ecology* **35**: 1–22.

Wiegand, T., Camarero, J.J., Ruger, N., and Gutierrez, E., 2006, Abrupt population changes in treeline ecotones along smooth gradients, *Journal of Ecology* **94**: 880–892.

Wiegand, T., Jeltsch, F., Hanski, I., and Grimm, V., 2003, Using pattern-oriented modeling for revealing hidden information: a key for reconciling ecological theory and application, *Oikos* **100**: 209–222.

Zadeh, L.A., 1965, Fuzzy sets, *Information and Control* **8**: 338–353.

Zadeh, L.A., 1975, The concept of a linguistic variable and its application to approximate reasoning – 1, *Information Sciences* **8**: 199–249.

ISSUES AND CHALLENGES OF INCORPORATING FUZZY SETS IN ECOLOGICAL MODELING

VINCENT B. ROBINSON
University of Toronto, 3359 Mississauga Road North, Mississauga, ON L5L 1C6 Canada

Abstract. An information-based framework is presented for spatially explicit GIS-based ecological modeling. Within this framework some of the important issues and challenges of incorporating fuzzy sets in spatially explicit population models (SEPM) are discussed. Examples of current work are used to illustrate the main issues and challenges facing the incorporation of fuzzy sets in ecological modeling. Among the challenges to be discussed are fuzzy-based techniques for data acquisition, model control/ evaluation, heterogeneous representations of spatial data, parameterization of models, and hypothesis testing. There is special attention given to the issue of habitat modeling and presence/absence problem. Many scientific issues facing the incorporation of ecological models will be raised such as hypothesis testing, and relationship between statistical analysis and fuzzy techniques. Software availability is discussed as challenge for past and future. use of fuzzy techniques directed at handling uncertainty in GIS-based SEPM.

Keywords: fuzzy sets, spatially explicit population model, geographic information system, habitat model, ecological geographic variables

1. Introduction

Changes in the landscape of either natural or human cause can significantly influence the distribution of species over space and through time. Spatially explicit population models (SEPMs) have developed to the level of simulating how animals may disperse across a landscape, hence affecting the metapopulation dynamics of a species (Lurz et al., 2001). These models explicitly link landscape information with species specific ecological knowledge. The spatially explicit nature of SEPMs makes them especially attractive for use in developing policies, strategies, and tactics for a wide range of environmental protection and security issues. Indeed, they are

A. Morris and S. Kokhan (eds.), Geographic Uncertainty in Environmental Security, 33–51.

valued in environmental protection/conservation efforts precisely because of their utility in contrasting future management scenarios (Liu et al., 1995).

It is the inherent fuzziness of biological organization that has prompted some to apply fuzzy techniques in ecological studies (Schaefer and Wilson, 2002). It is also accepted now that fuzzy techniques can be used to model, or manage, uncertainties inherent in GIS databases and processing (Robinson, 2003). Precisely because of their spatially explicit nature, these models have been increasingly used in conjunction with geographic information systems (GIS). The modeling of bird or mammal metapopulation dynamics generally requires data representing the landscape and the population dynamics of particular species of interest. Thus, this chapter first presents an information-based framework for spatially explicit GIS-based ecological modeling. Within this framework an overview of how and where fuzzy sets may be used to discuss strategies for incorporating fuzzy information processing in SEPM efforts. Among the challenges discussed are fuzzy-based techniques for data acquisition, model control, heterogeneous representations of spatial data, parameterization of models, and hypothesis testing. The relationship between fuzzy techniques and statistical methods in ecological modeling is briefly addressed. The challenge of adopting fuzzy-based techniques in the ecological modeling field(s) will also be discussed. Many scientific issues facing the incorporation of ecological models will be raised along with those of cost and software availability.

In its presentation of the issues and challenges of incorporating fuzzy techniques in SEPM, this chapter will tend to focus upon bird and small mammal species. Birds have been used for millennia as indicators of environmental health or degradation. In addition, they may carry pathogens such as avian flu. Small mammals are among the most easily studied mammals, especially in relation to landscape attributes. Like birds, they also can carry pathogens such as rabies and plague. Thus, these two broad categories of biological entities are of particular relevance to the issue of environmental protection and security due to their responses to landscape level changes and their importance as potential vectors of human pathogens.

2. Information Theoretic Framework

Lima and Zollner (1996) discuss information-based approaches to modeling the movement and dispersal of animals. These approaches can be considered to represent a continuum that at one extreme is characterized by theoretical studies assuming animals disperse in random directions for random distances settling in the nearest detectable habitat patch(s). Such studies rarely use actual landscapes, hence are of limited value for management, or protection purposes. At the other extreme are approaches that attribute considerable cognitive abilities to animals that include the use of spatial memory and learning. These models often make use of techniques from the field of artificial intelligence to model. These also tend to be poorly matched with real world landscape data and often make use of theoretical landscapes. In between these two extremes are those approaches that are intermediate in spatial scale where animals are typically given knowledge only about their nearby landscape and have no information about greater landscape. The model is often based on animals moving in the direction of the greatest detectable resource abundance or dispersing in direction of best detectable living site.

These approaches can be loosely categorized as being population- or individual-based models. As depicted in Figure 1, population-based models typically make use of a maximum dispersal distance to determine which habitat patches are potential destinations for individuals that need to be relocated from a patch with a surplus population to a patch of habitat that can accommodate a greater population density. Hence, the relationship is actually between the spatial entities of patches. The allocations are usually performed using random draws. Often there is little regard for obstacles to movement found in an intervening landscape matrix. In contrast, individual-based models base the relocation of individuals on distribution of habitat,

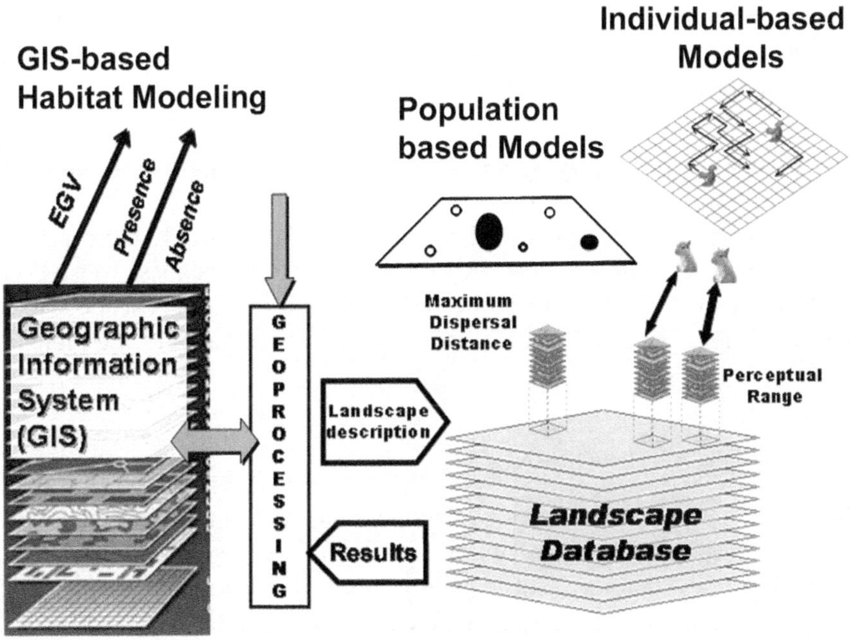

Figure 1. Conceptual overview of the relationship between GIS-based landscape modeling, population-based models of movement, and individual-based models of movement

and other landscape features, that are within the perceptual range of an individual (or social group). Hence the spatial relationship is more directly represented as being between an animal and landscape features.

Figure 1 also depicts in a conceptual manner some of the commonalities found in most all information-based modeling approaches. Information-based approaches use some version of perceptual range. In the case of population level models it is equivalent to the use of maximum dispersal distance. However complex or simplistic, they use rules of movement that control the movement of animal entities. They rely on a standard of plausibility (Lima and Zollner, 1996). The major reason for relying on a standard of plausibility is both the lack of spatially explicit behavior information on species' movements and difficulty of gathering such information (Zollner and Lima, 2005).

The process of going from ecogeographic variables (EGV) to arrive at a spatially explicit habitat model remains an important phase in a GIS-based approach. In addition, the behavioral and population model parameters contribute to the interpretation and execution of the rules of movement in concert with landscape data. At each step in this process there are questions of uncertainty to be addressed. Robinson (2002) discusses in detail the role fuzzy information processing may play at each step in this process. He notes the inherent fuzziness of the geographic information that forms the basis for the information used by the ecological model. In addition, he shows how fuzziness is inherent in the basic parameters of the models of dispersal/movement.

3. Incorporating Fuzzy Sets in GIS-Based Landscape Representations

In the context of the framework summarized in Figure 1, there are two major topics to consider when preparing to use a GIS database to support a SEPM exercise. First, there is the representation of the landscape without specific reference to habitat. Second, of crucial importance is the modeling of habitat.

3.1. LANDSCAPE INFORMATION

There are a number of sources of fuzziness inherent in the construction of a GIS database that may, or may not, subsequently be used for SEPM. Many GIS databases that are subsequently used for landscape level SEPM purposes were not originally assembled for that specific purpose. Figure 2 summarizes some the main input streams to a GIS database by the type of inputs. One of the major input streams is from sensor data. Land cover layers are typically used to reclassify the landscape into habitat versus nonhabitat patches; and remote sensing inputs are a common input upon which those classifications are made. The fuzzy classification of and representation of remote sensing data is a well-studied problem, yet its potential use and/or effect on subsequent use in landscape level studies is less well studied (Arnot et al., 2004). Techniques such as fuzzy c-means (or k-means) clustering or related techniques can be used to produce a classification of land cover that preserves the inherent fuzziness that characterizes such representations (Comber et al., 2005; Robinson 1988).

Another major input stream is from expert opinion (Figure 2). In the context of these types of ecological models, a good example of such data would be bird count data collected by experienced bird surveyors. As noted later this is inherently fuzzy. Expert opinion on land cover and other environmental variables may also be characterized as linguistic variables (Bordogna et al., 2006; Zhu et al., 2001). Sketch maps of some form are not uncommon. Even in the case of bird counts there is a kind of sketch map technique whereby the observer uses a bull's-eye form to approximate the location of a bird in terms of distance and direction. Other sketch maps depicting routes have been shown to be amenable to analysis using F-histograms and linguistic descriptions (Skubic et al., 2004).

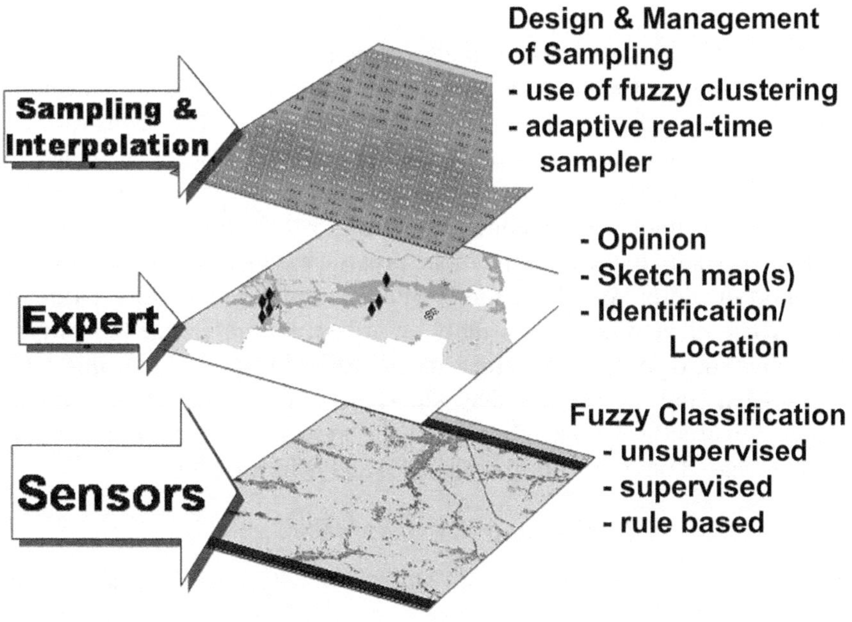

Figure 2. Major input streams used to create GIS database for landscape level studies and some associated fuzzy techniques

The third major input stream is associated many conventional methodologies used to create GIS-database layers from measurements on the ground based on some spatially explicit sampling scheme then interpolated to fill in the space to create a complete layer. One of the more common layers created in this fashion is the digital elevation model (DEM). Others include phenomena like soils, vegetation, wetlands, etc. In this context fuzzy clustering has been used to construct a sampling strategy from remotely sensed data or other reconnaissance information then interpolation methods used to generate a nonfuzzy representation such as soil characteristics (Odeh et al., 1990). It is less common for explicitly fuzzy techniques to be used for spatial interpolation (Dragicevic, 2005), especially in connection with efforts focused on SEPM. In contrast to this use of fuzzy sets, Graniero and Robinson (2003) have shown how a real-time adaptive sampler using fuzzy logic can increase the accuracy of such data collection while simultaneously lowering the cost.

3.2. HABITAT MODELING

Commonly, GIS-based models of animal movement require spatially explicit representation of the distribution of habitat as well as nonhabitat landscape elements. Uncertainty in the construction of a habitat model can have many different sources. It is also clear that in many cases defining a crisp boundary between habitat and nonhabitat is not ecologically meaningful, nor realistic. For example, Dettmers and Bart (1999) constructed a Boolean logic rule-based model of worm-eating warblers in included the condition that location have a slope between 35–44. However, does this mean that a location with a slope of 44.5 or 35.1 should be considered not habitat? Typically theories of the niche provide the theoretical underpinning to support the definition of suitability of a location as habitat for a particular species. The niche concept is usually described with reference to similarity of an optimal, or best, state. Thus, most habitat models rely on this logic. This has led some to suggest that whenever similarity-based concepts are used to construct a habitat model, that the model construction and testing should be based on using fuzzy logic (Hill and Binford, 2002).

3.2.1. *Fuzziness of Habitat Suitability Index (HSI)*

One common methodology is based on the Habitat Suitability Index (HSI). Burgman et al. (2001) point out several possible sources of uncertainty in the construction of a typical HSI. They emphasize the reliance, due to limited data, on expert opinion. They proceed to demonstrate how to use fuzzy numbers to encapsulate the uncertainty of the relationship between the value of an EGV (e.g., distance to forest) and level of suitability (i.e., ranges from 0 to 1). Reliability bounds for the best estimates of suitability were shown to vary significantly as a function of the characteristics of individual patches of landscape. Their example using the Florida scrub-jay illustrates how dependent some avian habitat modeling exercises are on expert opinion.

Hill and Binford (2002) also discuss the inherent fuzziness of the HSI. They make special reference to the reliance on expert opinion. Although they do not develop the concept of linguistic variables (Zadeh, 1978), they do demonstrate how a HSI model can be constructed using ambiguous categories using the theory of fuzzy sets. However, their demonstration of its application is not a GIS-based modeling approach. This lack of a GIS-based methodology is typical of HSI development. Hence, HSI results can often be of limited value to landscape level modeling efforts.

3.2.2. *The presence/absence problem*

Although expert opinion is often the basis for HSI, there is considerable effort directed at obtaining presence (and/or absence) data. Avian studies often rely upon detection of individuals through sight or sound (Austen et al., 2001; Jenkins et al., 2003). Mammalian studies often use trapping, photography, or tracks to determine presence (Carroll et al., 1999; Oehler and Litvaitis, 1996; Orrock et al., 2000). In both cases it is often tacitly assumed, for modeling purposes, that lack of detection is tantamount to absence. Hence the prevalence of methods assuming presence/absence data (Fielding and Bell, 1997).

In their study of why sparrow distributions do not match the spatially explicit predictions of habitat models, Jenkins et al. (2003) noted there were Commission errors where some habitat does not contain birds. These are areas where prior events have depleted species numbers, but because the birds do not disperse long distances, they do not reoccupy it quickly. This is consistent with theories of metapopulation dynamics.

> Omission errors where some birds remain in unsuitable locations (i.e., nonhabitat or habitat of low quality) that were formerly suitable locations. This situation may result from the species having high site fidelity (Jenkins et al., 2003).

This has led to competing, nonfuzzy statistical methodologies for predicting the suitability of landscape elements as habitat (Brotons et al., 2004). Some of these methods include the use of pseudoabsence approaches, researchers are cautioned to be mindful of their study design as well as biases inherent in the presence data (Pearce and Boyce, 2006). Although this group of techniques makes use of expert opinion mostly as a general guide as to what EGV to include, recent results suggest that any more detailed inclusion of expert opinion may actually decrease the predictive power of the model (Pearce et al., 2001; Seoane et al., 2005). Such results call into question the uncritical inclusion of expert opinion.

In summary, an estimate of only the amount of habitat or only the bird population may present an overly optimistic view of a species plight. Hence, the fuzziness inherent in determination of what constitutes habitat through presence/nonpresence data can have significant implication for environmental protection and conservation. Note that habitat in this context is represented as a crisp, nonfuzzy concept when, in fact, it more accurately is represented as a fuzzy concept.

3.2.3. *Fuzzy habitat modeling with presence data*

Like many fields, fuzzy c-means (or, k-means) was one of the first methods used in ecological analysis. In this case it is used commonly to develop classifications of ecological communities (Equihua, 1990). It has since been used to characterize the gradations in landscape-related characteristics within areas where species such as bats (Medellin et al., 2000) and squirrels (Bridges, 2003) are present. In these cases, it has been shown to represent more faithfully the internal heterogeneity of the habitat, thus avoiding the

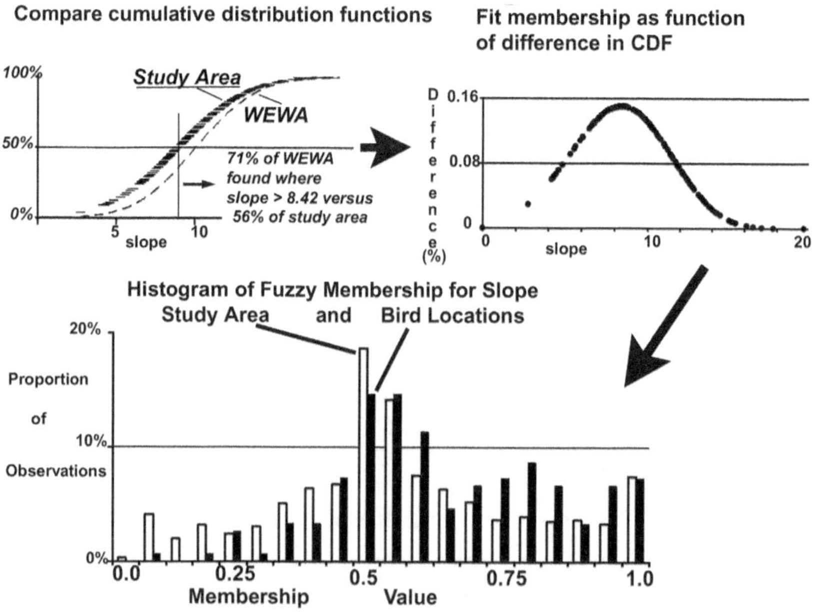

Figure 3. Example of an experiment inferring habitat rules by fuzzification of CDF-based methodology using presence data

oversimplification that occurs with crisp classifications. However, this approach has not been used to actually infer habitat from presence-only data.

One approach to inferring a fuzzy habitat model from presence-only data is currently under exploration (Hodgson, in progress). Figure 3 summarizes a portion of the process whereby cumulative distribution functions (CDFs) between presence data and the surrounding study area are compared to determine if there is any difference in the two. If there is a difference it is used in conjunction with membership function generation to determine if the variable can be used as an important EGV describing habitat for the species. In concept this is a fuzzification of the nonfuzzy approach described by Dettmers and Bart (1999). The end result would be a GIS raster layer in which each cell is coded according to the degree to which it is a member of the habitat set.

4. Fuzziness and Model Parameters

Attention has largely focused on the uncertainty concerning landscape level GIS information as input to support SEPM while few studies have investigated the effects of uncertain model parameters. One such study, using a crisp classification of habitat/nonhabitat landscapes found that errors in dispersal-mortality parameters resulted in greatest prediction errors, followed by mobility errors, and lastly landscape errors (Ruckelshaus et al., 1997). Results such as those in Ruckelshaus et al. (1997) suggest that focusing solely on the fuzziness of the landscape representations may not be as important to the reliability of model predictions as focusing on addressing the uncertainty in model parameters. Their results suggest that uncertainty surrounding dispersal parameters is a significant problem that should receive attention from those constructing SEPMs. Since crucial parameters in the dispersal/movement models concern decisions of where to move to and when to stop moving, it can be considered analogous to a control problem. Indeed Robinson (2002) has drawn parallels between control of mobile agents, robotic, and otherwise, to the problem of modeling individual animals in support of SEPM. Similarly, fuzzy cellular automata have been used to model insect infestation spread in a forested region of British Columbia (Bone et al., 2006).

In one GIS-based modeling effort, a fuzzy membership function was used to model the likelihood that a grid cell can be reached by a squirrel starting from a source patch. Results using the fuzzy approach were more consistent with field data (DeGenst et al., 2001). This suggests the potential utility of fuzzy sets in the parameterization of spatially explicit models at the population level model (see Figure 1).

Another approach is the modeling of animals at the individual level where vital parameters controlling dispersal/movement, such as perceptual range, that cannot be precisely derived from field and/or experimental work. For example, Mech and Zollner (2002) report the perceptual range of several common sciuridae as ranges, such as for the gray squirrel the perceptual range is between 300 m and 400 m. However it is not biologically meaningful to say 400.1 m is completely and utterly not perceived while a landscape feature at 400 m is perceived. Typically perceptual ranges are modeled as if they occur on isotropic surface rather than the anisotropic surfaces that exist in reality. It has been shown that such context dependent perceptual ranges can have an effect on the movement behavior of individuals (Olden et al., 2004).

Robinson and Graniero (2005) describe how an object-oriented probe mechanism can be used to incorporate fuzziness in an effective manner at the level of model parameters such as the perceptual range. Essentially the individual animals are simulated as a population of objects where each class has a different decision-making model of whether to move, where to move, and when to stop moving. However, each object does not directly draw its information from the GIS-based database, but rather through a probe mechanism (Graniero and Robinson, in press). Robinson and Graniero (2005) model squirrel dispersal using a fuzzy set representation of perceptual range. It is modeled as a function of distance on an isotropic surface. In

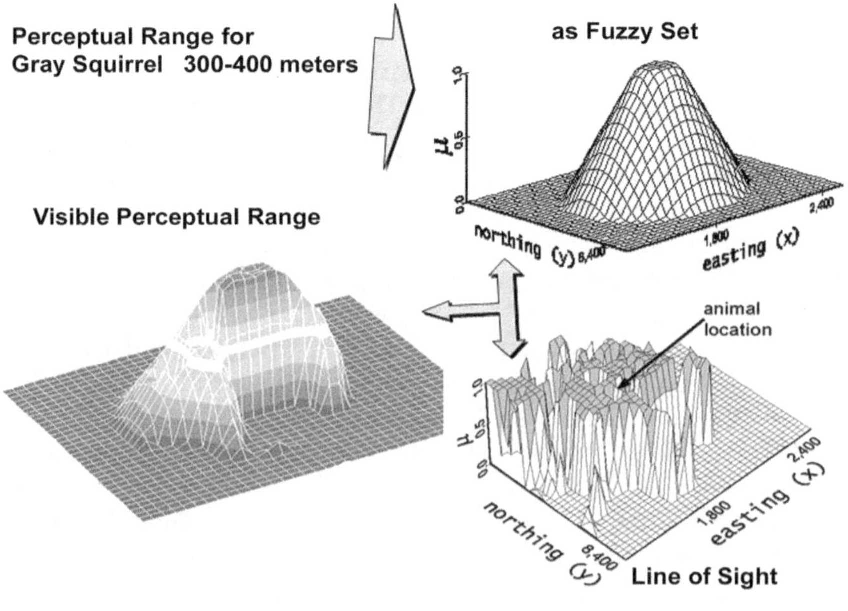

Figure 4. Derivation of context sensitive perceptual range as a fuzzy set

addition, a line of sight set is calculated to represent the degree to which landscape elements are visible to a squirrel (i.e., animal object) up in a tree to arrive at a fuzzy set of visible landscape. The combination of perceptual range and visible landscape provided a set of landscape elements who were members to some degree in the visible perceptual range represented as a fuzzy set (see Figure 4). This joint fuzzy representation of perceptual range and visibility formed an individual object's view of the landscape database derived from the GIS layers. The landscape information was then used in a fuzzy-based decision-making model of that had two submodels. The movement submodel concerned the locational choice for the next move. Once the individual object had been moved to that location, the residence submodel was used to determine whether or not it would stay in the location. In other words, did that location meet the requirements for establishing a home range that would sustain an individual? Details of these submodels can be found in Robinson and Graniero (2005).

Since fuzzy set theory is particularly rich in set connectives, four decision models were used to simulate dispersal behavior. Three were based on fuzzy set theory using compensatory, noncompensatory, and Yager style connectives. For comparison, a fourth model was a crisp (nonfuzzy) model. Hence there were four populations (i.e., classes) of animal objects with each population using a different decision model. Figure 5 illustrates how the different decision models yield differing spatially explicit behavior. Thus, this approach illustrated the ability of a fuzzy-based approach to represent variations in decision making as might occur within populations. Since success at finding a home range location is critical to support modeling of metapopulation dynamics and species spread, this result illustrates how fuzziness incorporated in a spatially explicit ecological model may help explain variations in rates of recolonization or dispersal.

Since squirrels tend to set up home ranges in areas not contested by conspecifics, simulations were for landscapes with different levels of conspecific coverage. The base line situation was the case of social fencing where there were no "holes" in the conspecific landscape. Variations in the base line were implemented by randomly allowing for 20%, 40%, and 60% percent of the landscape to be composed of unoccupied "holes." A conspecific "hole" was therefore a spatially explicit goal for a disperser. Results of these simulations showed that variations in landscape perception, and consequent use of the information in the decision submodels, did have an effect on the simulated animal's ability to find a suitable location for establishing a home range. Furthermore, in all situations the simulated animals using the fuzzy-based submodels were more successful at finding a home range than were those using a submodel based on crisp logic (Robinson and Graniero, 2005).

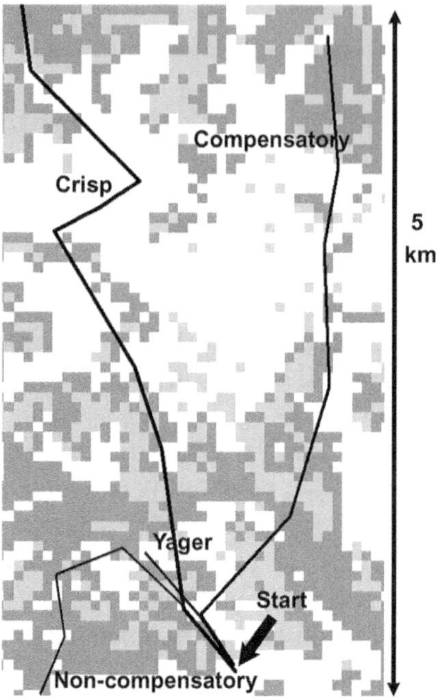

Figure 5. An example of the effect of differing decision models on spatial behavior of an individual given the same starting location and surrounding landscape. Note that the gray regions are deciduous forest while the white regions tend to be nonforest

The example shown in Figure 5 illustrates that a range of behaviors within a single species can be modeled. This is more realistic than assuming all animal agents have a single, identical decision submodels. In addition, the simulations yielded plausible results that do not rely upon stochasticity, thus providing a more satisfying framework for understanding how movement behavior may be influenced by landscape and ecological factors as well as the decision submodels. However, it remains to be determined whether or not such results would actually have significant impact on the ultimate demographic dynamics of a SEPM.

5. Hypothesis Testing

Although fuzzy techniques have been used to classify landscape elements, characterize model parameters, and characterize within habitat/community variations, they have rarely been used to directly test hypotheses. When discussing fuzzy techniques in conjunction with statistical analysis two primary issues arise:

Use of fuzzy techniques versus statistical analysis: In other words, can fuzzy techniques be used to appropriately accept or reject a hypothesis?

Use of fuzzy techniques along with statistical techniques: In other words, can the two approaches be used in a complementary fashion rather than as competing paradigms?

In the end, there is some question as to whether or not fuzzy techniques should, or even can, be evaluated against nonfuzzy techniques. One of the problems with constructing such an evaluation is the specification of a metric of evaluation that is meaningful to both.

5.1. FUZZY TECHNIQUES VERSUS STATISTICAL ANALYSIS

In habitat modeling it is common to see logistic regression used as if the assumption of presence/absence has been met. This, and other approaches, relies upon probability theory via the use of inferential statistics. As an alternative to this approach fuzzy-based techniques of evaluating habitat using presence data are beginning to emerge (Hill and Binford, 2002; Hodgson, in progress). For example, Hill and Binford (2002) present a fuzzy-based tests that allows habitat models to be accepted or rejected according to the degree to which they under/overestimate habitat potential. They suggest that their approach provides a means of not only using approximate reasoning, but also a means of providing defensible conclusions. In the context of ecological modeling culture, this issue remains a significant challenge for those researchers seeking to incorporate fuzzy sets in their modeling efforts.

Fuzzy clustering has been used to investigate the description of vegetation communities (Equihua, 1990) and spatial structure of animal populations (Feoli and Zerihun, 2000; Schaefer and Wilson, 2002). Schaefer and Wilson (2002) showed how fuzzy clustering can depict the ambiguous structure of caribou and walleye population's spatial structure. The advantage over the hard clustering technique was that it was able to convey the degree of uncertainty of belonging to a population cluster of both unambiguous as well as problematic individuals. Using both crisp and fuzzy clustering techniques Nicholls and Tudorancea (2001) demonstrated how fuzzy cluster analysis whereby species can be ranked according to their contribution to the structure of the dendrogram. Until this demonstration there had been few good methods for accomplishing this quantitatively (Nicholls and Tudorancea, 2001).

5.2. FUZZY TECHNIQUES WITH STATISTICAL TECHNIQUES

As suggested by Smithson (2005), it seems reasonable to consider that the joint, complementary use of fuzzy and statistical techniques as a powerful way of testing hypotheses. Equihua (1990) used analysis of variance to compare the results of fuzzy clustering with a crisp approach. In an analysis of bat habitat in a disturbed tropical forest environment, Medellin et al. (2000) used generalized linear models to regress bat community variables against a fuzzy membership score derived from fuzzy clustering. However, the statistical analysis of fuzzy membership patterns remains relatively unknown in GIS-based ecological modeling even though there are now techniques out that will explicitly test for spatially explicit patterns (Hagen-Zanker et al., 2005).

Rather than using statistical analysis simply as an analysis of the outcome of a fuzzy-based exercise, it has been demonstrated in other fields that it can used to choose fuzzy membership functions and form fuzzy rules Focusing on the problem of uncertain parameter estimation in a hydroecological model, Mackay et al. (2003) illustrate the use, a two stage methodology, where in the first stage Monte Carlo simulations are run. Then each simulation result is evaluated by regressing simulated evaporative fraction from a regional hydroecological model and surface temperature. For each regression, the coefficient of determination (R^2) is used as a fuzzy measure of the goodness of fit for its respective simulation result. Hence the fuzzy set is composed of the set of R^2 measures for all simulations, to which a fuzzy information-theoretic tool is applied to form a restricted set in which only good simulations are retained. A restricted set is used as an ensemble solution in the second stage of parameter estimation. This illustrates the potential for a combined statistical-fuzzy approach to address the issue of uncertainty in model parameters. However, it has yet to be developed in a convincing manner in the context of SEPM.

6. Software

The cost of developing and implementing fuzzy computational tools may have been, in the past, an obstacle to incorporating fuzzy sets in GIS-based ecological modeling. To some extent, it remains an issue since it is still relatively rare for a commercially available GIS software system to support fuzzy information processing. A notable exception is the IDRISI GIS software package (Eastman, 1999). However, there are add-on software works, not commercially available, that are coupled with commercially available GIS products such as ESRI's products (Benedikt et al., 2002; Yanar and Akyurek, 2006). However, to date these tools have not yet had substantial presence the ecological modeling literature. Other tools, outside GIS software packages, have been developed.

Although the code for the fuzzy c-means algorithm was published in 1984 (Bezdek et al., 1984), fuzzy clustering remains a relatively infrequently used technique in ecological modeling. Shortly after Bezdek et al.'s (1984) paper appeared, applications in the classification of land cover using remote sensing data did appear (Fisher and Pathirana, 1994; Robinson and Thongs, 1986). Even though it is now a technique well known to remote sensing specialists, the additional information afforded by a fuzzy classification of land cover is generally not exploited by GIS-based ecological modeling efforts. The availability of software such as the FuzME (Minasny et al., 2002) package have supported the use of fuzzy clustering in ecological modeling (Bridges, 2003; Schaefer and Wilson, 2002). In addition to FuzME, the fuzzy c-means and FANNY techniques reported in Nicholls and Tudorancea (2001) are now freely available as part of the R statistical project (Dimitriadouet et al., 2005; Maechler, 2005). The experience of fuzzy clustering illustrates that the cost and availability of software to support fuzzy techniques in GIS-based ecological modeling is a very real issue. The challenge is to not only make these techniques available but also to illustrate their use, utility, and availability to the ecological modeling community at large.

7. Concluding Comments

Fuzzy clustering techniques have been used sporadically in habitat studies that form the basis for inferring habitat patches from a GIS database. The results of using fuzzy clustering techniques suggest that they preserve, and represent, more faithfully the spatial variability of habitat than do the more traditional crisp techniques. This is of particular importance when incorporating rare community types in habitat analysis since crisp techniques tend to overlook, or eliminate, their contribution to habitat. Thus, one of the main issues and challenges of incorporating fuzzy sets in ecological modeling is the degree to which the results are ecologically meaningful, supportive, or not, of a given hypothesis, and ecologically plausible.

Perhaps more importantly is the modeling process itself. The degree to which the model and its parameterization have plausible ecological interpretations is a challenge. Like most applications in the ecological literature, this chapter's discussion has exclusively considered type 1 fuzzy sets. Since uncertainty is the fundamental reason that models incorporate fuzzy sets, it is important to consider the fact that the fuzzy membership functions are themselves subject to uncertainty. Such uncertainty is modeled using type 2 fuzzy sets (Fisher et al., in press). It would be useful for the purposes of constructing fuzzy ecological models to have an appreciation of whether or not, the added computational complexity and burden of type 2 fuzzy sets would result in any significant improvement of the models and their performance.

This chapter has but scratched the surface in presenting a variety of issues and challenges regarding the incorporation of fuzzy sets in GIS-based ecological modeling. The topics are an intersection of research and applications in geographic information science, behavioral ecology, landscape ecology, remote sensing, fuzzy systems, and database theory.

Acknowledgments

The author gratefully acknowledges the partial support of Discovery Grant RGPIN 44611–02 from the Natural Sciences and Engineering Research Council (NSERC) of Canada. The partial support of NATO to support the author's participation in the workshop held in Kyiv, Ukraine is most gratefully acknowledged.

References

Arnot, C., Fisher, P. F., Wadsworth, R., and Wellens, J., 2004, Landscape metrics with ecotones: pattern under uncertainty, *Landscape Ecology* **19**:181–195.

Austen, M. J., Francis, C. M., Burke, D. M., and Bradstreet, M. S. W., 2001, Landscape context and fragmentation effects on forest birds in Southern Ontario, *The Condor* **103**:701–714.

Benedikt, J., Reinberg, S., and Riedl, L., 2002, A GIS application to enhance cell-based information modeling, *Information Sciences* **142**:151–160.

Bezdek, J. C., Ehrlich, R., and Full, W., 1984, FCM: the fuzzy c-means clustering algorithm, *Computers and Geosciences* **10**:191–203.

Bone, C., Dragicevic, S., and Roberts, A., 2006, A fuzzy-constrained cellular automata model of forest insect infestations, *Ecological Modelling* **192**:107–125.

Bordogna, G., Chiesa, S., and Geneletti, D., 2006, Linguistic modelling of imperfect spatial information as a basis for simplifying spatial analysis, *Information Sciences* **176**:366–389.

Bridges, L. M., 2003, *Spatial Scale and Environmental Structure: Habitat Selection of Adult Grey Squirrels (Sciurus carolinensis) in Central Ontario*, Master's thesis, Trent University.

Brotons, L., Thuiller, W., Araujo, M. B., and Hirzel, A. H., 2004, Presence-absence versus presence-only modelling methods for predicting bird habitat suitability, *Ecography* **27**:437–448.

Burgman, M. A., Breininger, B. W., Duncan, B. W., and Ferson, S., 2001, Setting reliability bound on habitat suitability indices, *Ecological Applications* **11**:70–78.

Carroll, C., Zielinski, W. J., and Noss, R. F., 1999, Using presence-absence data to build and test spatial habitat models for the Fisher in the Klamath Region, U.S.A., *Conservation Biology* **13**:1344–1359.

Comber, A., Fisher, P., and Wadsworth, R., 2005, What is land cover? *Environment and Planning B: Planning and Design* **32**:199–209.

DeGenst, A., Canters, F., and Gulink, H., 2001, Uncertainty modeling in buffer operations applied to connectivity analysis, *Transactions in GIS* **5**:305–326.

Dettmers, R. and Bart, J., 1999, A GIS modeling method applied to predicting forest songbird habitat, *Ecological Applications* **9**:152–163.

Dimitriadou, E., Hornik, K., Leisch, F., Meyer, D., and Weingessel, A., 2005, *The e1071 Package*, R Project for Statistical Computing, http://cran.r-project.org/.

Dragicevic, S., 2005, Multi-dimensional interpolations with fuzzy sets, In: *Fuzzy Modeling with Spatial Information for Geographic Problems*, Petry, F. E., Robinson, V. B., and Cobb, M. A., eds., Springer, Berlin, pp. 143–158.

Eastman, J. R., 1999, *Idrisi32 Guide to GIS and Image Processing*, Clark Labs, Worcester, MA.

Equihua, M., 1990, Fuzzy clustering of ecological data, *Journal of Ecology* **78**:519–534.

Feoli, E. and Zerihun, W., 2000, Fuzzy set analysis of the Ethiopian rift valley vegetation in relation to anthropogenic influences, *Plant Ecology* **147**:219–225.

Fielding, A. H. and Bell, J. F., 1997, A review of methods for assessment of prediction errors in conservation presence/absence models, *Environmental Conservation* **24**:38–49.

Fisher, P. F. and Pathirana, S., 1994, The evaluation of fuzzy membership of land cover classes in the suburban zone, *Remote Sensing of Environment* **34**:121–132.

Fisher, P., Cheng, T., and Wood, J., in press, Higher order vaguess in geographical information: empirical geographical population of type-n fuzzy sets, *GeoInformatica* **29**:106–128.

Graniero, P. A. and Robinson, V. B., 2003, A real-time adaptive sampling method for field mapping in patchy, heterogeneous environments, *Transactions in GIS* **7**:31–54.

Graniero, P. A. and Robinson, V. B., in press, A probe mechanism to couple spatially explicit agents and landscape models in an integrated modelling framework, *International Journal of Geographical Information Science*

Hagen-Zanker, A., Straatman, B., and Uljee, I., 2005, Further developments of a fuzzy set map comparison approach, *International Journal of Geographical Information Science* **19**:769–785.

Hill, K. E. and Binford, M. W., 2002, The role of category definition in habitat models: practical and logical limitations of using Boolean, indexed, probabilistic, and fuzzy categories, In: *Predicting Species Occurrences: Issues of Accuracy and Scale,* Scott, J. M., Heglund, P. J., Morrison, M. L., Haufer, J. B., Raphael, M. G., Wall, W. A., and Samson, F. B., eds., Island Press, Washington, DC, pp. 97–106.

Hodgson, P., in progress, *An Exploratory Study in the Use of Fuzzy Set to Model the Habitat of Worm-eating Warblers in the Land Between the Lakes National Recreation Area (tentative title)*, Master's thesis, University of Toronto.

Jenkins, C. N., Powell, R. D., Bass, O. L., Jr., and Pimm, S. L., 2003, Why sparrow distributions do not match model predictions, *Animal Conservation* **6**:39–46.

Lima, S. L. and Zollner, P. A., 1996, Towards a behavioral ecology of ecological landscapes, *Trends in Ecology and Evolution* **11**:131–135.

Liu, J., Dunning, J. B., Jr., and Pulliam, H. R., 1995, Potential effects of a forest management plan on Bachman's sparrows (*Aimophila aestivalis*): linking a spatially explicit model with GIS, *Conservation Biology* **9**:62–75.

Lurz, P. W. W., Rushton, S. P., Wauters, L. A., Bertolino, S., Currado, I., Massoglio, P., and Shirley, M. D. F., 2001, Predicting grey squirrel expansion in North Italy: a spatially explicit modelling approach, *Landscape Ecology* **16**:407–420.

Mackay, D. S., Samanta, S., Ahl, D. E., Ewers, B. E., Gower, S. T., and Burrows, S. N., 2003, Automated parameterization of land surface process models using fuzzy logic, *Transactions in GIS* 7:139–153.

Maechler, M., 2005, *The cluster Package*, R Project for Statistical Computing, http://cran.r-project.org/.

Mech, S. G. and Zollner, P. A., 2002, Using body size to predict perceptual range, *Oikos* **98**:47–52.

Medellin, R. A., Equihua, M., and Amin, M. A., 2000, Bat diversity and abundance as indicators of disturbance in neotropical rainforests, *Conservation Biology* **14**:1666–1675.

Minasny, B. and McBratney, A. B., 2002, FuzME version 3.5, Australian Centre for Precision Agriculture, The University of Sydney, Asutralia, December 12, 2003.

Nicholls, K. H. and Tudorancea, C., 2001, Application of fuzzy cluster analysis to Lake Simcoe crustacean zooplankton community structure, *Canadian Journal of Fisheries and Aquatic Sciences* **58**:231–240.

Odeh, I. O. A., McBratney, A. B., and Chittleborough, D. J., 1990, Design of optimal sample spacings for mapping soil using fuzzy k-means and regionalized variable theory, *Geoderma* **47**:93–112.

Oehler, J. D. and Litvaitis, J. A., 1996, The role of spatial scale in understanding responses of medium-sized carnivores, *Canadian Journal of Zoology* **74**:2070–2079.

Olden, J. D., Schooley, R. L., Monroe, J. B., and Poff, N. L., 2004, Context-dependent perceptual ranges and their relevance to animal movements in landscapes, *Journal of Animal Ecology* **73**:1190–1194.

Orrock, J. L., Pagels, J. F., McShea, W. J., and Harper, E. K., 2000, Predicting presence and abundance of a small mammal species: the effect of scale and resolution, *Ecological Applications* **10**:1356–1366.

Pearce, J. L., Cherry, K., Drielsma, M., Ferrier, S., and Whish, G., 2001, Incorporating expert opinion and fine-scale vegetation mapping into statistical models of faunal distribution, *Journal of Applied Ecology* **38**:412–424.

Pearce, J. L. and Boyce, M. S., 2006, Modelling distribution and abundance with presence-only data, *Journal of Applied Ecology* **43**:405–412.

Robinson, V. B., 1988, Some implications of fuzzy set theory applied to geographic databases, *Computers, Environment, and Urban Systems* **12**:89–97.

Robinson, V. B., 2002, Using fuzzy spatial relations to control movement behavior of mobile objects in spatially explicit ecological models, In: *Applying Soft Computing in Defining Spatial Relations*, Matsakis, P. and Sztandera, L. M., eds., Physica-Verlag, Heidelberg, pp. 158–178.

Robinson, V. B., 2003, A perspective on the fundamentals of fuzzy sets and their use in geographic information systems, *Transactions in GIS* 7:3–30.

Robinson, V. B. and Graniero, P. A., 2005, Spatially explicit individual-based ecological modeling with mobile fuzzy agents, In: *Fuzzy Modeling with Spatial Information for Geographic Problems*, Petry, F. E., Robinson, V. B., and Cobb, M. A., ed., Springer, Heidelberg, pp. 299–334.

Robinson, V. B. and Thongs, D., 1986, Fuzzy set theory applied to the mixed pixel problem of multispectral landcover databases, In: *Geographic Information Systems in Government*, Opitz, B. K., ed., A. Deepak Publishing, Hampton, VA, pp. 871–885.

Ruckelshaus, M., Hartway, C., and Kareiva, P., 1997, Assessing the data requirements of spatially explicit dispersal models, *Conservation Biology* **11**:1298–1306.

Schaefer, J. A. and Wilson, C. C., 2002, A fuzzy structure of populations, *Canadian Journal of Zoology* 2235–2241.

Seoane, J., Bustamante, J., and Diaz-Delgado, R., 2005, Effect of expert opinion on the predictive abiliyt of environmental models of bird distribution, *Conservation Biology* **19**:512–522.

Skubic, M., Blisard, S., Bailey, C., Adams, J. A., and Matsakis, P., 2004, Qualitative analysis of sketched route maps: translating a sketch into linguistic descriptions, *IEEE Transactions on Systems, Man, and Cybernetics* **34**:1275–1282.

Smithson, M., 2005, Fuzzy set inclusion: linking fuzzy set methods with mainstream techniques, *Sociological Methods and Research* **33**:431–461.

Yanar, T. A. and Akyurek, Z., 2006, The enhancement of the cell-based GIS analyses with fuzzy processing capabilities, *Information Sciences* **176**:1067–1085.

Zadeh, L. A., 1978, Fuzzy sets as a basis for a theory of possibility, *Fuzzy Sets and Systems* **1**:3–28.

Zhu, A. X., Hudson, B., Burt, J., Lubich, K., and Simonson, D., 2001, Soil mapping using GIS, expert knowledge, and fuzzy logic, *Soil Science Society of America Journal* **65**:1463–1472.

Zollner, P. A. and Lima, S. L., 2005, Behavioral tradeoffs when dispersing across a patchy landscape, *Oikos* **108**:219–230.

RELIABILITY OF VEGETATION COMMUNITY INFORMATION DERIVED USING DECORANA ORDINATION AND FUZZY *C*-MEANS CLUSTERING

L. BASTIN*
School of Engineering and Applied Science, University of Aston, Birmingham B4 7ET, United Kingdom.

P.F. FISHER
School of Informatics, City University, London, EC1V 0HB, United Kingdom

M.C. BACON
Department of Geography, University of Leicester, Leicester, LE1 7RH, United Kingdom

C.N.W. ARNOT
Department of Geography, University of Leicester, Leicester, LE1 7RH, United Kingdom

M.J. HUGHES
Environmental Change Research Centre, University College London, 26 Bedford Way, WC1H 0AP, United Kingdom

Abstract. Descriptions of vegetation communities are often based on vague semantic terms describing species presence and dominance. For this reason, some researchers advocate the use of fuzzy sets in the statistical classification of plant species data into communities. In this study, spatially referenced vegetation abundance values collected from Greek *phrygana* were analysed by ordination (DECORANA), and classified on the resulting axes using fuzzy *c*-means to yield a point data-set representing local memberships in characteristic plant communities. The fuzzy clusters matched vegetation communities noted in the field, which tended to grade into one another, rather than occupying discrete patches. The fuzzy set representation of the community exploited the strengths of detrended correspondence analysis while retaining richer information than a TWINSPAN classification of the same data. Thus, in the absence of phytosociological benchmarks, meaningful and manageable *habitat* information could be

*To whom correspondence should be addressed. Lucy Bastin, School of Engineering and Applied Science, University of Aston, Birmingham B4 7ET, UK. Email: l.bastin@aston.ac.uk

A. Morris and S. Kokhan (eds.), Geographic Uncertainty in Environmental Security, 53–74.

derived from complex, multivariate *species* data. We also analysed the influence of the reliability of different surveyors' field observations by multiple sampling at a selected sample location. We show that the impact of surveyor error was more severe in the Boolean than the fuzzy classification.

Keywords: phrygana, multivariate analysis, error, uncertainty, ecotone, fuzzy *c*-means, DECORANA, TWINSPAN

1. Introduction

Ecological surveys often require the collection and storage of spatially referenced multivariate data on plant and animal occurrences (Jongman et al., 1995). For any one sample location, there can be a large number of records on presence/absence or abundance of a variety of different species. To quote Gauch (1982b), such data "are complex, showing noise, redundancy, internal relations and outliers". This encourages the use of multivariate techniques such as correspondence analysis to condense ecological data so that it can be interpreted in terms of meaningful communities. For example, in the case of UK plant data, multivariate species data at a sample point can be characterized into defined assemblages such as "limestone pavement" or "oak/beech woodland" (Rodwell, 1991, 1992). Such an interpretation can be derived using some, or all, of the following three techniques (Gauch, 1982b; Jongman et al., 1995);

Ordination projects the multidimensional data into a smaller number of dimensions, which have been identified as important by statistical patterns in the data.

Classification labels the samples as belonging to particular defined communities, usually by some form of clustering or divisive analysis.

Interpretation results are investigated in more detail, in terms of expected communities, or with reference to underlying physical and climatic phenomena measured at the same locations.

The end product is a data-set showing the occurrence of recognized vegetation associations in a study area. When these data are spatially referenced, they can be used to illustrate characteristic patterns of vegetation change across the landscape.

The ultimate aim of the survey reported here was to produce numerical vegetation information at spatially referenced sample points, which could be combined with multispectral data for co-kriging, to produce an interpolated vegetation map covering the whole study area. Our aim was to handle the semantic and spatial fuzziness of semi-natural vegetation, by modelling vegetation communities as fuzzy sets, where sample locations could have memberships ranging between 0 (no similarity) and 1 (full, typical membership). A previous study (Eqihua, 1990) has explicitly compared fuzzy clusters generated from reciprocal averaging ordination axes to a hierarchical TWINSPAN classification of the same data. The specific purpose of this paper is to perform a similar comparison on plant data from another geographical source, and to extend the comparison to the handling of survey variation and error.

1.1. BACKGROUND

In ecology, the ordination process is often carried out using detrended correspondence analysis, as implemented in the DECORANA programme (Hill, 1979a; Hill and Gauch, 1982), in order to allow interpretation of species data in terms of vegetation communities. DECORANA is specifically designed to isolate meaningful, uncorrelated axes from multidimensional data clouds, thus summarizing underlying covariance and co-occurrences, and reducing redundancy. Most importantly in the context of this paper, the ordination process is intended to reduce the impact of sampling and surveyor variation by "selectively deferring noise to late axes" (Gauch, 1982a).

Classification of samples into community categories is commonly carried out by hierarchical divisive clustering, for example, using TWINSPAN (Hill, 1979b). Classified results may then be interpreted by comparison to existing, recognized communities or as new community definitions in their own right. For the purposes of this study, we needed to classify vegetation point samples into meaningful communities. There were two important considerations in this classification; semantic fuzziness (vagueness in the definition of a community; Moraczewski 1993a, b), and spatial fuzziness (the tendency for vegetation units to mix and grade into one another; Fisher, 2000; Foody, 1992, 1996).

1.2. FUZZY SETS

It is not necessary to articulate the full details of fuzzy sets here, as these are more than adequately covered elsewhere in the popular (Kosko, 1993), the technical (Klir and Yuan, 1995; Kruse et al., 1994), and the ecological literature (Dale, 1988; Roberts, 1986, 1989a, b). In summary, however, given an exhaustive classification of a group of samples with an approach based in traditional crisp, hard or Boolean sets, any case must belong to one and only one class. This "membership" is coded by the integers 1 or 0, indicating that the case either completely belongs to the class or does not. In a classification based in fuzzy sets this condition is relaxed, so that any one case may belong *to some degree* to all classes. Thus "fuzzy set membership" is measured as a real number in the range between 1 and 0, where 1 indicates that the case totally belongs to that class (and usually has no affinity with any other class) and where 0 indicates that the case does not have any affinity with that class. It is usual, but not obligatory, that fuzzy memberships arising from a classification are normalized, that is to say the sum of the fuzzy memberships for any particular case in all classes is equal to 1. Associated with the fuzzy set membership is a formal logic (Klir and Yuan, 1995), but consideration of that is beyond the concern of this chapter.

1.3. SEMANTIC FUZZINESS

The definition of a vegetation "community" is dependent on the scope and amount of detail in the study, and on the pre-existing data for an area. Such a definition could be as broad as "tundra" (CORINE, European Environment Agency, 1991) or as specific as the basophyllous maquis scrub association *Phillyrea latifolia-Arbutetum unedonis* – (Loidi et al., 1994). Communities and assemblages can be formally parameterized, on the basis of extensive survey and subsequent statistical analysis. An example of such a numerical classification is the National Vegetation Classification scheme for UK vegetation (Rodwell, 1991, 1992, 1995). However, communities may, as in the case of phytosociological groupings, be described in semantic terms (e.g., "Community A: species X dominant, with occasional species Y and infrequent stands of species Z") (Roberts, 1996). Such descriptions are particularly suited to representation using fuzzy set theory (Moraszewski, 1993a, b).

In general, the definition of a community involves the identification of one or more fundamental "indicator" species which are always present, and a variety of other species whose occurrence and frequency are used to refine the final classification. The definition of a community on the basis of species presence/absence alone is particularly well suited to final classifications based on sequential hierarchical splitting (e.g., TWINSPAN). In terms of species abundance, the use of "pseudospecies" offers some opportunity to represent within TWINSPAN the fact that the same species may be important to several communities, but occur at different characteristic frequencies. However, the final classification is still hard or Boolean, with a sample being either a member or not a member of a community grouping. In this paper we specifically tackle an alternative method – fuzzy classification – which allows each sample to be a member of *one or more* communities, and measures the strength of each membership as a number ranging between 0 and 1, where 1 represents perfect typicality, and 0 represents no similarity to that cluster. We also examine the sensitivity of the classification to variations in identification.

Where a numerically parameterized vegetation classification exists (e.g., the UK's National Vegetation Classification, or NVC), a fuzzy interpretation of field data may be attempted by interpreting "similarity metrics", as though they were "membership values". For example, the segregation of vegetation samples to NVC classes using TABLEFIT yields a variety of different indices, showing how similar each sample is to its final label category (Table 1). However, no such baseline numerical classification was available for the *phrygana* vegetation types described in this paper. For the purposes of this trial study, vegetation classes were derived from the available data itself; this runs the risk of circularity, and this problem is more fully discussed later in the chapter.

TABLE 1. Results of fitting a vegetation sample to NVC classes, using TABLEFIT. The sample is assigned to five possible categories (labelled 1st–5th), in order of closeness of fit. The closeness of fit is determined by indices A to E. These represent: A – a weighted average of B, C, D and E, used as an overall goodness of fit: B – Compositional satisfaction: C – Mean constancy of species in the sample: D – Dominance satisfaction : E – Dominance constancy. All these terms are further explained in the supporting documentation for TABLEFIT (Hill, 1996; Hill, 1989)

Sample Q3	*1	Parameters = Nolich	Cover%	Sp &c					
	CORINE	NVC	A	B	C	D	E		
1s	C31.	H 1e	69	93	53	100	64	Calluna-Fest	Species-
t	2251							ovin heath	poor
2n	C31.	H 1b	68	64	75	100	65	Calluna-Fest	Hyp phy-
d	2251							ovin heath	Cla imp
3r	C31.	H 9c	58	43	69	89	71	Calluna-Desc	Species-
d	2254							flex heath	poor
4t	C31.	H13a	56	32	87	92	76	Calluna-Clad	Cla arb-
h	2257							arbu heath	Cla ran
5t	C31.	H 9e	56	40	61	82	81	Calluna-Desc	Molinia
h	2254							flex heath	caerul

1.4. SPATIAL FUZZINESS

Plant community units ("stands") are rarely separated by clear boundaries, but may grade or blend into one another (Foody, 1992, 1996; Moraszewski, 1993a). Thus, at any point in space, the assemblage of species present can show similarities to several defined communities. Figure 1 shows data on three plant species, collected from five evenly spaced quadrats. The cover proportion of each species can be seen to change fairly evenly along the sequence, from wet grassland to dry heath. Using a "hard" community classification, it would be difficult to define the status of the intermediate vegetation samples in the sequence, even if the two end points could be satisfactorily classified as pure vegetation types.

This general phenomenon of mixing and gradation is difficult to represent with the hard categories of a TWINSPAN classification; an alternative method is to use fuzzy clustering to extend and "soften" the boundaries of the defined communities. Legendre and Legendre (1998, p. 371), briefly mention such fuzzy clustering techniques, but state that "these models have not been used in ecology yet". In fact, some practical use had at that time been made of fuzzy clusters as vegetation community descriptors (Eqihua, 1990; McCracken, 1994; Zhang, 1994; Zhang and Oxley, 1994), and Eqihua (1990) concluded that, "In comparison to TWINSPAN, it [fuzzy clustering] seems able to produce clusters which are more strongly correlated with relevant external variables". Eqihua makes several compelling arguments for the philosophical shift to representing vegetation communities as fuzzy sets. Of these, perhaps the most interesting is the observation that, even when communities are seen as discrete entities, those entities are identified post hoc from sampled data; therefore some mixing is inevitable, since field samples cannot be selectively placed in "pure" locations. This study adopts the classification strategy of Eqihua (1990) and Zhang and Oxley (1994); namely, DECORANA ordination to concentrate meaningful variation on early axes, which then define the variable space for a fuzzy clustering procedure.

Fuzzy clustering does in general seem natural and attractive as a tool for interpreting ecological data. Unimodal species response functions along environmental gradients appear almost as ready-made fuzzy set functions, and more complex data dispersed in a multidimensional coenospace will always display some degree of overlap at the margins of the more obvious clusters. Fuzzy ordination of field data against known or estimated environmental gradients is becoming more common (e.g., Boyce, 1998; Andreucci et al., 2000), with easy-to-use packages such as "Fuzzy Grouping" (© Pisces Conservation Ltd.) also offering a fuzzy *c*-means clustering option. Fuzzy clustering in *species*, rather than *environmental* space is less common, (though Brown, 1998 bases a supervised classification

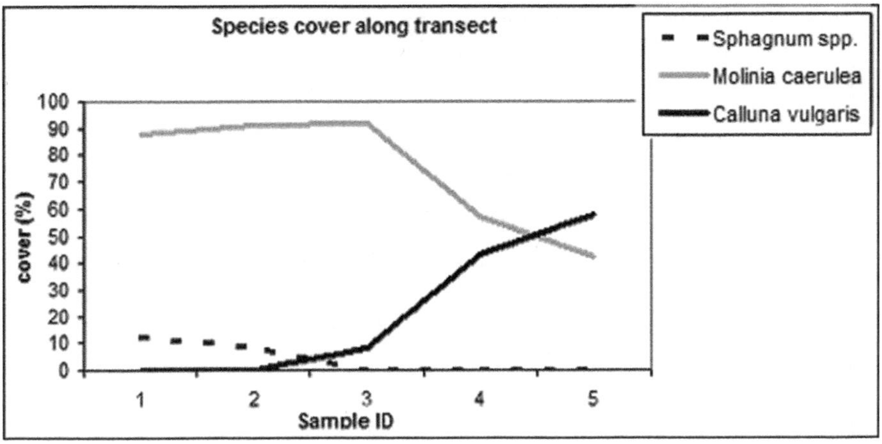

Figure 1. Percentage cover of three plant species can be seen to change relatively smoothly along a sample transect spanning a wet-grassland/dry heath transition

on forest classes pre-defined using expert knowledge on species abundance), and it is this approach which will be discussed in this chapter.

2. Methods

2.1. DATA COLLECTION

The data described in this paper were collected to provide a ground record of an area of limestone *phrygana* near Thessaloniki, in northern Greece (*22.93N, 40.74 E*: see Figure 2). *Phrygana* constitutes a Greek group of *garrigue* communities, characterised by sclerophyllous dwarf shrubs on dry soils, *Phrygana* is generally thought to be a secondary vegetation community (resulting from human degradation of high pine and oak forest, or natural succession on abandoned fields) which may be very old, but whose natural succession to sclerophyllous woodland is restricted by domestic grazing (Diamantopoulos et al., 1994; Bergmeier et al., 1996).

This particular study area was dominated by *Cistus incanus* and *Quercus coccifera* (kermes oak), and shrub density varied across the area, ranging from open areas of grassland and bare soil to a denser, patchy form of tall scrub, known as *shiblyak*. Kermes oak shrublands of this type cover around 1,500 km^2 in Greece, and are of considerable economic importance for livestock grazing (Tsiouvaras et al., 1999). Vegetation samples were taken at 190 locations (see Figure 3), over a period of 12 days. The semi-natural vegetation formed a mosaic between more homogeneous areas of cultivated wheat and pine plantations, which were digitized but not surveyed for this study.

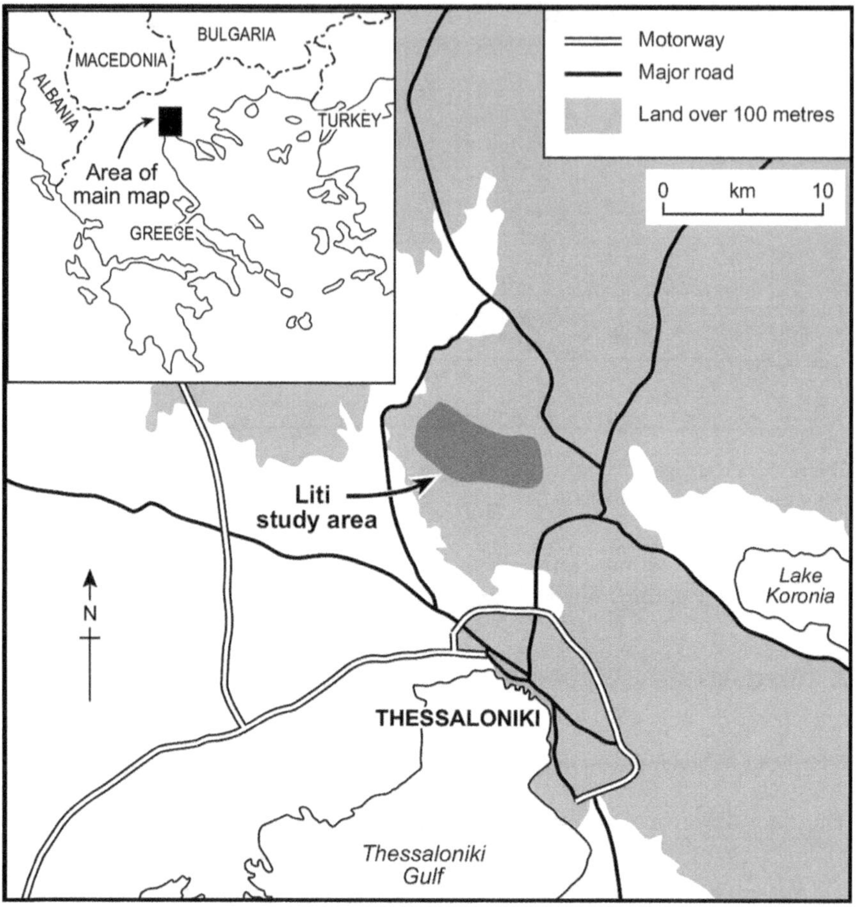

Figure 2. The location of the study area near Thessaloniki in northern Greece

At each location, a 2m×2m quadrat and a 20 m line transect sample were taken, in order to record vegetation at two different scales. The quadrat was surveyed in detail, recording all species and ground covers, and their estimated percentage cover. Each transect recorded distances covered by shrub species, trees and patches of open ground, thus representing the vegetation on a coarser scale. Sample locations were recorded with Garmin hand-held GPS (averaging for 10–15 min). For verification, ten of these points were also located using Ashtech differential GPS, and the maximum locational error was found to be 32 m in any direction.

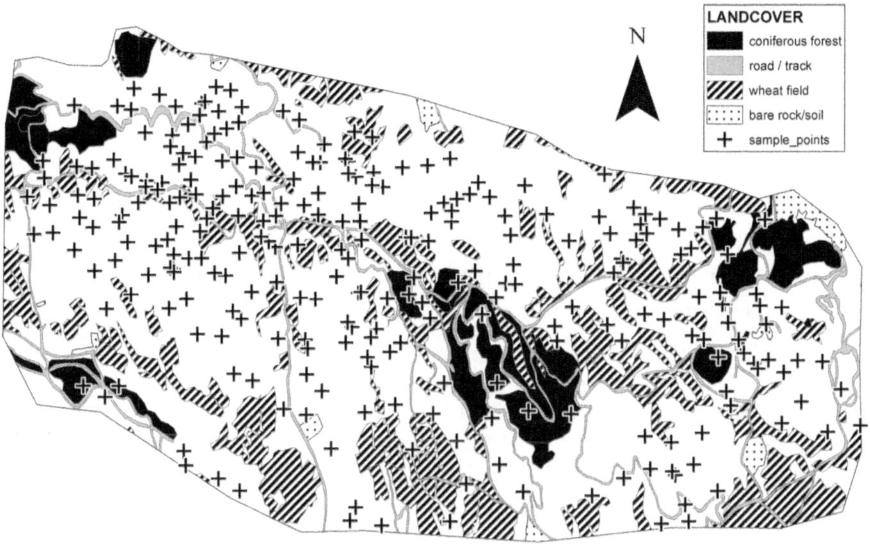

Figure 3. The location of sampling points within the study area

2.2. ANALYSIS OF THE 190 SAMPLES

The data collected were analysed by a combination of ordination using DECORANA (Hill, 1979a; Hill and Gauch, 1980) and fuzzy *c*-means classification (Bezdek et al., 1984). For comparative purposes they were also subject to conventional classification using TWINSPAN (Hill, 1979b).

2.2.1. *Ordination*

Quadrat and transect data represented, respectively, the finely structured grass/herb community and the coarser matrix of grassland, scrub, bare ground and trees. Analysis for both followed the same protocol (described below), but only the transect data are discussed in this chapter. First, a DECORANA ordination was carried out, yielding a matrix of axis scores, shown in Table 2. Rare species were downweighted.

2.2.2. *Classification*

The DECORANA axis scores were used as the basis for fuzzy classification to four categories. The Fuzzy *c*-means (fcm) algorithm (Bezdek et al., 1984) was used, with a Euclidean norm. After some experimentation, a fuzziness exponent (m) of 1.8 was used.

TABLE 2. DECORANA species scores for transect data

Species/ground cover	Axis 1	Axis 2	Axis 3	Axis 4
Asparagus acutifolius	324	15	–13	103
Astragalus massiliensis	365	365	287	0
Bare ground/grasses	210	157	164	120
Cistus incanus	270	24	–60	25
Cornus sanguinea	0	133	223	114
Crataegus monogyna	185	90	329	232
Lonicera periclymenum	343	0	19	145
Paliuris spina-christi	239	209	210	–188
Phillyrea angustifolia	127	221	–104	–101
Pinus halepensis	110	107	214	187
Prunus spinosa	233	231	528	491
Pyrus pyraster	286	17	128	82
Quercus coccifera	167	119	78	63
Quercus spp. (deciduous)	92	230	42	145
Rosa spp.	122	137	273	217
Rubus spp.	156	188	69	114
Stone	368	0	251	287

2.2.3. *Test replicates – handling of surveyor error and uncertainty*

The field surveyors for this study had varying amounts of experience in vegetation identification and field survey. It was therefore important to assess the variation between samples which was due to individual surveyors (e.g., differences in species identification and measurement), rather than to genuine variation in vegetation. Over the period of the study, as new plant species were identified, they were added to a common list, and named according to a consensus among the surveyors. Thus, even if a plant species could not be identified accurately to species level, it was defined by the same name by all surveyors. In addition, surveyors worked in pairs but the pairing of individuals was rotated over the study period to minimize bias. However, surveyor variation could still be important, and in order to quantify its effects, nine different pairs of surveyors recorded the same standard transect.

Replicate transects were broadly very similar; the same species were identified by all surveyors, though at different frequencies (Figure 4).

For fuzzy classification, each of the nine replicate samples was treated, in turn, as the 191st sample, producing nine separate and alternative data-sets

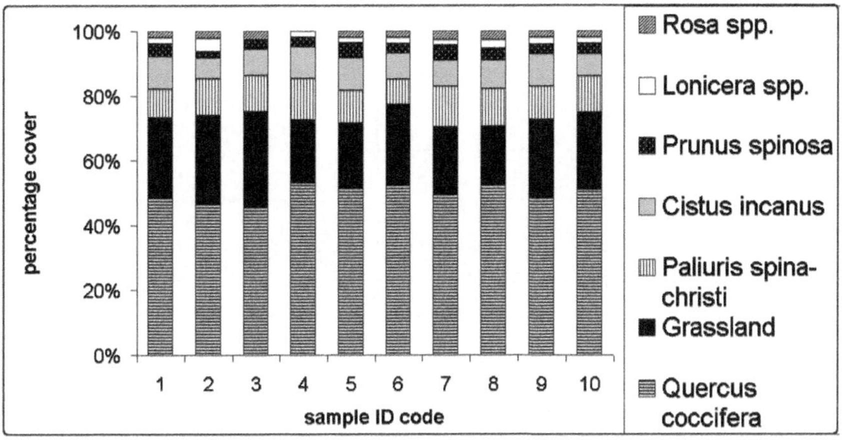

Figure 4. Transect composition as defined by different pairs of surveyors

of 191 samples. In each of these datasets, the final sample had been recorded by a different surveyor, but all other samples were identical. These data-sets were subjected to DECORANA ordination, and the resulting axis scores used for fuzzy *c*-means clustering, to produce fuzzy membership values for all 191 samples for each of the nine data-sets.

In TWINSPAN, the data-sets were categorized to the third level of division, (with default pseudospecies cuts) to give a group number for every sample. In this case, all nine samples were added to the original 190 at once, so that the segregated groups could be meaningfully compared.

3. Results

Figure 5 illustrates the influence of two of the four DECORANA axes on the TWINSPAN clustering of transect data points. When these data points are mapped to their relative locations in space. the different TWINSPAN categories can be seen to have clear affinities and relationships to one another (Figure 6). The TWINSPAN classification therefore is useful for segregating the vegetation data into communities. However, any sample point has a hard (Boolean) relationship to the resulting communities, being a member of one and only one category. The fuzzy clustering shows somewhat similar spatial variations in class memberships (Figure 7) but recognises four classes, as compared to eight in the TWINSPAN results.

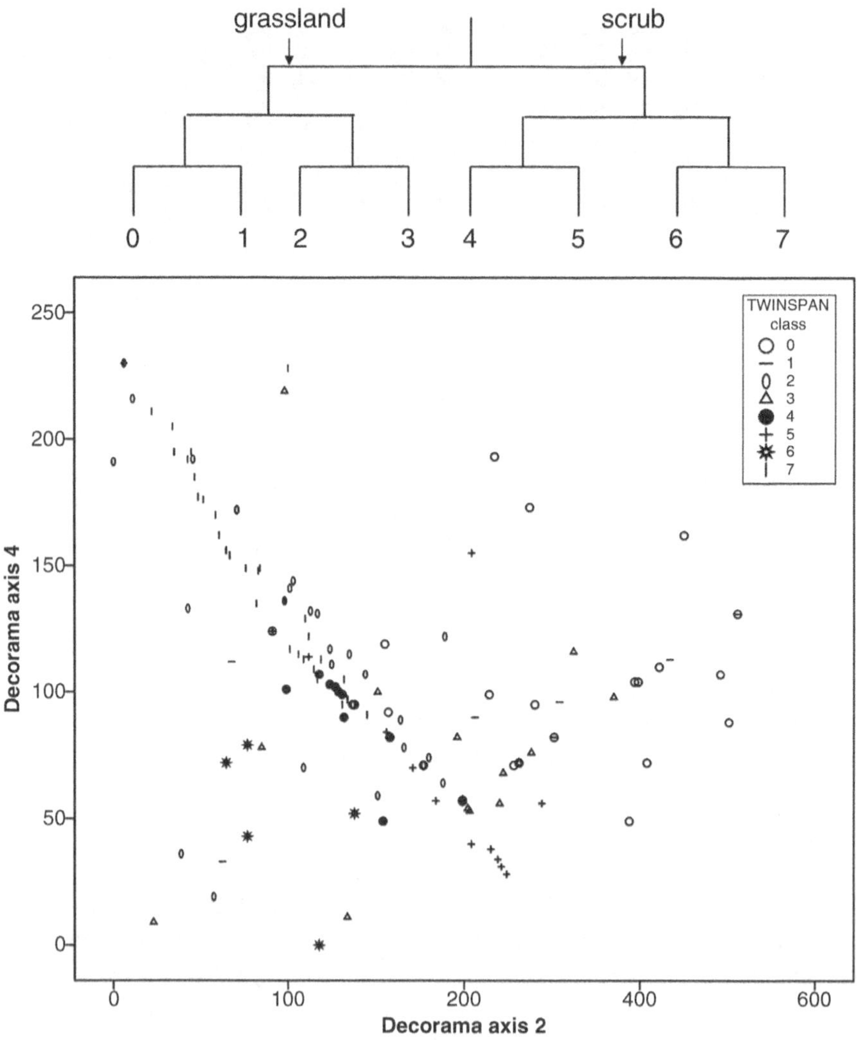

Figure 5. Categories produced by a 3-level TWINSPAN classification of transect data, plotted against DECORANA axes 2 and 4. The tree view shows the relationship between classes in the TWINSPAN hierarchy, and the rough split of classes between grass and scrub communities

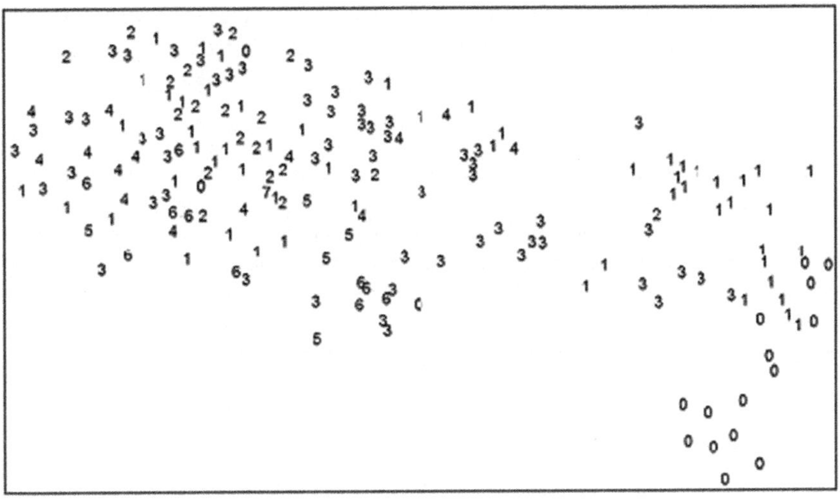

Figure 6. TWINSPAN-classified transect samples mapped according to their relative geographical positions. Number labelling is as for Figure 5

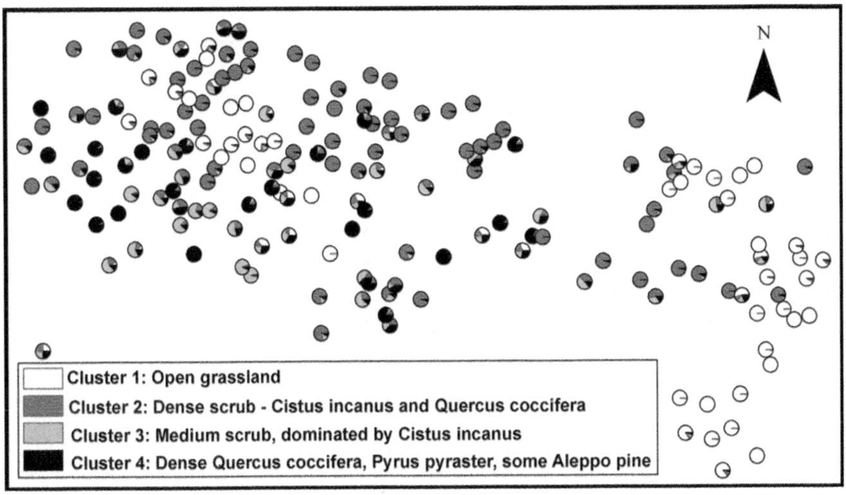

Figure 7. Transect memberships, mapped according to their relative geographical positions

The clusters corresponded to broad vegetation categories observed within the sample areas, with some significant relationships between some clusters and variables such as vegetation patchiness, shrub cover and texture[1] (see Table 3).

TABLE 3. Pearson product-moment correlation coefficients between the fuzzy memberships and various other summary vegetation indices. Indices are as follows: Open%, Low%, High% and Tree% = estimated percentage of open ground, low scrub, high scrub and trees at the site, Patch. = patchiness, Slope = slope angle of transect, in degrees. Measures of aspect such as "Northness" (cos(aspect)) showed very low linear correlations with fuzzy memberships

Cluster	OPEN%	LOW%	HIGH%	TREE%	Patch.	Texture	Slope
1	0.539	–0.267	–0.455	–0.081	–0.292	–0.329	0.119
2	–0.734	–0.008	0.862	0.024	–0.064	0.245	–0.039
3	–0.392	0.695	0.072	0.053	0.399	0.237	0.132
4	0.264	–0.175	–0.248	0.035	0.108	0.006	0.122

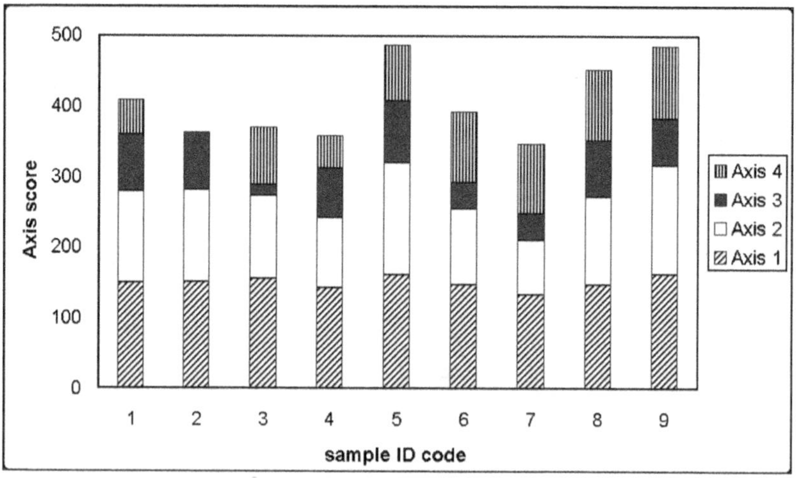

Figure 8. DECORANA axis scores for replicate transects

[1] Transect *texture* measured the total sum of all changes in vegetation height within the transect. Transect *patchiness* recorded the number of changes in broad vegetation type (e.g. shrub/tree/grass) along the transect.

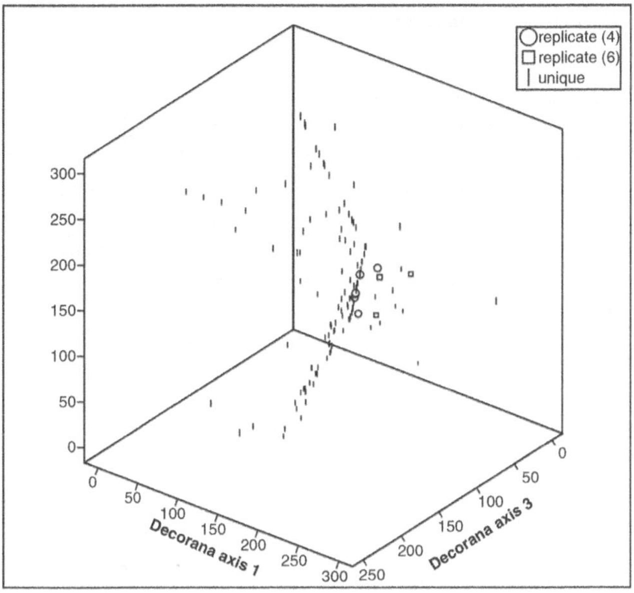

Figure 9. Replicate transects (and all unique transects) plotted against the first three axes of the DECORANA ordination. The nine replicate points are labelled according to whether TWINSPAN allocates them to class 4 or 6

The species recorded in the nine replicate transects, and their relative ground cover, were fairly consistent. Although DECORANA ordination did somewhat accentuate the differences between these replicate samples (Figure 8), they remain fairly close together in the ordination space, when viewed in the context of all the sample data (see Figure 9).

The results of the fcm classification (Figure 10) show a range of fuzzy memberships in three of the four possible clusters. All nine samples are dominated by Cluster 3 (which was positively correlated with medium shrub cover). However, the actual memberships varied between 0.6 and 1.0, and cluster 1 (open grassland) also contributed up to 38% of the total membership. Interestingly, the balance between clusters 1 and 3 was hardly reflected in the TWINSPAN classification, which divided the nine replicate samples between two classes at the second level (classes 4 and 6, as illustrated in Figure 6). This division appeared most strongly to reflect the membership in cluster 2 (dense *Cistus* scrub). The fuzzy classification retains valuable information in that, if "hardened" to a single class, it would produce a consistent classification (Cluster 3) for all replicates, but the three samples whose cover estimates were distinctly different (which were separated out completely by TWINSPAN) can still be identified by the strength of their membership in Cluster 2.

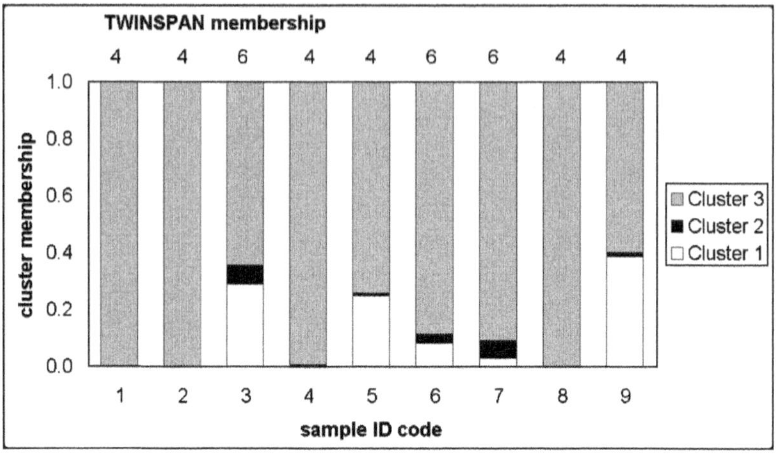

Figure 10. Comparison of the TWINSPAN and fuzzy classifications of the transect replicates

4. Discussion and Conclusions

The fuzzy classification approach, like TWINSPAN, offered a useful means of characterizing observed vegetation patterns and associations in the absence of a pre-existing subjective or numerical classification scheme. The fuzzy memberships, however, also allow modeling of vaguely defined community types, as well as helping to represent the spatial fuzziness observed at many natural vegetation boundaries. The fuzzy set representation of the community exploited the strengths of detrended correspondence analysis (e.g., handling of "packing" and non-linear responses, and downweighting of rare species) while retaining richer information than a traditional Boolean classification such as that generated by TWINSPAN.

The correspondences between the TWINSPAN and fuzzy classifications of repeated samples are also encouraging. Variations in recorded species composition were obvious in the replicate samples, but the effect of this type of uncertainty is often difficult to quantify. Both the "hard" and the fuzzy classification method identified the same samples as particularly different from the others. This implies that, in cases where continuously valued vegetation data are necessary, a fuzzy classification such as this, (a) has the power to discriminate between relatively similar samples, (b) allows partial membership in vegetation clusters to be quantified and used for further analysis, retaining more information from the original multi-species data-set, and (c) because of avoiding the hard decision of belonging to classes and replacing it with degree of membership is more robust to surveyor variance.

Most of the differences observed between the TWINSPAN and the fuzzy results stem directly from the natures of the two classifiers. TWINSPAN, as a divisive strategy, necessarily hinges on the identification of indicator species whose abundance values can be used as "breakpoints" between any number of clusters. While the classification produces discrete, "hard" clusters, they are generated using specifically ecological reasoning, and acknowledged assumptions about the ways in which species distributions are affected by environmental gradients. The fuzzy method, on the other hand, is constrained to produce the number of clusters requested, which means some experimentation and validation is necessary. While the fuzzy *c*-means clustering method in itself does not model ecological patterns such as co-occurrence, the authors believe that this element is added to the process by the use of the DECORANA package to project the data along more ecologically meaningful axes prior to clustering. Both methods, as used here, are limited by the data, and will find unique, natural centres of gravity for the specific data-set. This said, the correspondence observed between fuzzy cluster membership and a (limited) set of recorded variables (Table 3) is encouraging. The logical extension of this approach, therefore, is to calibrate the eigenvalues used for *c*-means clustering against a range of exemplar samples for defined plant communities, and expert botanical survey knowledge, to achieve a numerical classification similar to the UK's NVC. Alternatively, and again using expert knowledge, a set of fuzzy memberships for each species in each community type may be constructed, as in Brown (1998) The communities thus parameterized would effectively act as the "compositional nodes" described by Roberts (1989a): real or hypothetical templates to which the similarity of other samples is measured.

Several commentators have criticized the "fishing" approach, where fuzzy *c*-means and other techniques are used to pull out clusters from a cloud of data with no prior hypothesis, and no reference to external variables. However, this approach has a long heritage in cluster analysis, and could be argued to be more acceptable in the current era of data mining. In order to obtain any meaningful results from a fuzzy representation of vegetation classes, some clear aims and ground rules do need to be laid out a priori. The study described here highlighted several issues, as follows:

Axes must be well defined, and preferably have some ecological meaning. The example described here is not benchmarked to any numerical consensus classification, but relies on patterns of species dominance and consistency derived from the dataset itself. As discussed above, an alternative approach, which some researchers find preferable, is to make use of existing numerical classifications where they exist. No usable numerical system on which to base a supervised classification existed at the time of this study, and so the axes derived here would be difficult to generalize to other communities, or to multi-temporal contexts.

Desired number of clusters should ideally be defined a priori, and relate to the state of knowledge of the vegetation, BUT it must be recognized that an unsupervised classification may discriminate entirely different assemblages. The common practice for fuzzy classification is to select a number of "good" clusters, using a criterion such as the partition entropy. However, Dale (1988) asks with some justification "why, if you accept the need for a fuzzy solution, the desirable solution should be chosen as that which is most crisp!".

The m-value (fuzziness) of the classification should be experimentally determined and well justified. Varying the "fuzziness" of the classifier can lead to dramatically different results. Elsewhere in this volume, Fisher and Arnot note the variety of *m*-values used in the literature, and Dale (1988) complains that fuzzy approaches "disappoint in their sensitivity to the choice of exponent". The effect on the replicate transects of varying *m* and cluster number is shown in Figure 11. By application of a range of values of *m*, it is possible to generate richly informative "Type 2 fuzzy sets" (Fisher and Arnot [ibid.]). However, this approach tends to generate huge quantities of data, and there may be occasions when, for pragmatic analysis, a single *m*-value is required. With a known level of surveyor error and uncertainty (based on replicate surveys) and a desired number of clusters, we could select an fuzziness parameter which allows memberships to be hardened so that replicates KNOWN to represent the same area are consistent, while retaining the maximum amount of information on the variation between surveyors. Thus the clustering is calibrated to a set of surveyors with a known variance. For a four-cluster solution in this example, this threshold between discriminatory value and oversensitivity occurred at an *m*-value of 1.8.

By following these guidelines, it should be possible to represent, analyse and model fuzzy communities, rather than "blurry artifacts" (Palmer, 1994).

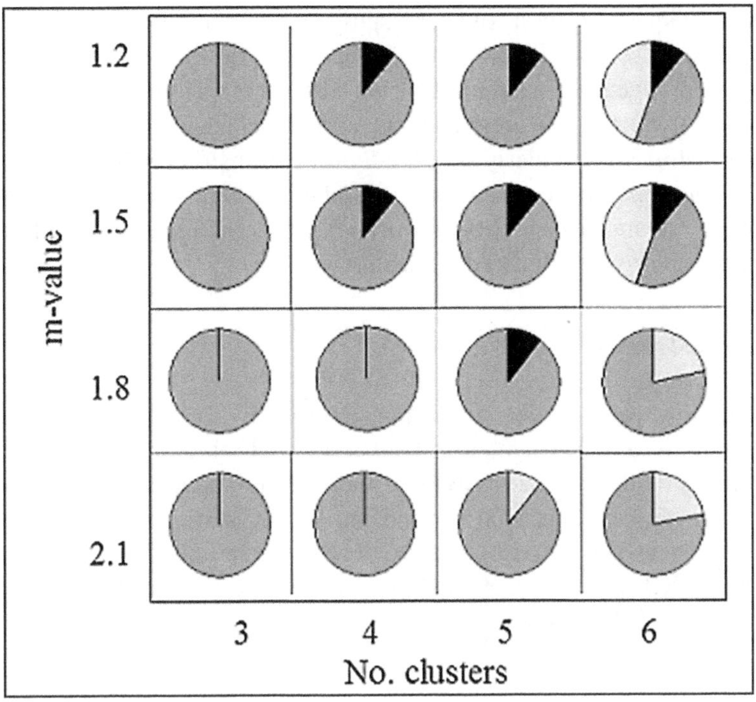

Figure 11. Hardened classifications for the nine replicate transects, for a variety of *m*-values and cluster numbers. On the left, all nine samples are dominated by membership in a single cluster, while at the centre of the matrix, one sample has majority membership in a different cluster. NB: because the classification is unsupervised, and cluster centres shift for each case, the three clusters seen here do not exactly correspond to the clusters discussed earlier, *except* for the case where cluster number = 4, and $m = 1.8$

In the case of this study, continuous data on vegetation classes were required for interpolation by co-kriging, and there are several other application areas where categorical data cannot be easily used; for example, in the modelling of vegetation succession over time, or of ecotones. A combination of ordination and fuzzy clustering was found to be a useful tool, not least because it is straighforward to analyse the membership strengths and their correlation with other continuous environmental variables, such as slope or precipitation. The presence/absence information yielded by the TWINSPAN clustering can also be related to such supplementary variables by methods such as logistic regression, whose use in an explicitly spatial context has been discussed by Smith (1994). However, the problem remains that with a hard classification, relatively subtle shifts in dominance between species can cause an apparently abrupt jump from one cluster to another. This is a particular problem in the analysis of spatial pattern, since the statistical tests and indices which can be applied to a set of single-valued points are limited. In addition, subtle changes in vegetation composition may be very meaningful in the context of environmental pressures such as climate change and over or under-grazing. This is particularly true for those semi-natural habitats such as the Greek *phrygana* or heath and wet grassland, where management (by fire, grazing or clearance) consists of artificially preventing succession to mature forest, scrub or wet woodland respectively.

Acknowledgements

The authors would like to thank Professor N. Silleos and Mr. A. Konstadinidis at the Aristotelean University of Thessaloniki for their help in organizing field work and providing facilities to support this work. Thanks also go to the students who formed the survey party for the field season. This work was supported by the EU Environment and Climate Programme under DG XII (contract number ENV4-CY96-0305), and was conducted as part of the FLIERS Project.

References

Andreucci, F., Biondi, E., Feoli, E., and Zuccarello, V. (2000) Modelling environmental responses of plant associations by fuzzy set theory. *Community Ecology*, **1**: 73–80.

Bergmeier, E. and Matthas, U. (1996) Quantitative studies of phenology and early effects of non-grazing in Cretan phrygana vegetation. *Journal of Vegetation Science*, **7**: 229–236.

Bezdek, J.C., Ehrlich, R., and Full, W. (1984) FCM: the fuzzy c-means clustering algorithm. *Computers and Geosciences*, **10**: 191–203.

Boyce, R.L. (1998) Fuzzy set ordination along an elevation gradient on a mountain in Vermont, USA. *Journal of Vegetation Science*, **9**: 191–200.

Brown, D.G. (1998) Mapping historical forest types in Baraga County Michigan, USA as fuzzy sets. *Plant Ecology*, **134**: 97–111.

Dale, M.B. (1988) Some fuzzy approaches to phytosociology: ideals and instances. *Folia Geobotanica et Phytotaxonomica*, **23**: 239–274.

Diamantopoulos, J., Pirintsos, S.A., Margaris, N.S., and Stamou, G.P. (1994) Variation in Greek phrygana vegetation in relation to soil and climate. *Journal of Vegetation Science*, **5**: 355–360.

European Environment Agency (1991) CORINE Biotopes. Published by the Commission of the European Communities, Directorate-General Environment, Nuclear Safety and Civil Protection Office for Official Publications of the European Communities.

Equihua, M. (1990) Fuzzy clustering of ecological data. *Journal of Ecology*, **78**: 519–534.

Fisher, P.F. (2000) Sorites Paradox and Vague Geographies. *Fuzzy Sets and Systems*, **113**: 7–18.

Foody, G.M. (1992) A fuzzy-sets approach to the representation of vegetation continua from remotely sensed data – an example from lowland heath. *Photogrammetric Engineering and Remote Sensing*, **58**: 221–225.

Foody, G.M. (1996) Fuzzy modelling of vegetation from remotely sensed imagery. *Ecological Modelling*, **85**: 3–12.

Gauch, H.G. (1982a) Noise reduction by eigenvector ordinations. *Ecology*, **63**: 1643–1649.

Gauch, H.G. (1982b) *Multivariate Analysis in Community Ecology*. Cambridge University Press, Cambridge.

Hill, M.O. (1979a) DECORANA – A FORTRAN program for detrended correspondence analysis and reciprocal averaging. *Ecology and Systematics*. Cornell University, Ithaca, NY.

Hill, M.O. (1979b) TWINSPAN – A FORTRAN program for arranging multivariate data in an ordered two-way table by classification of the individuals and attributes. *Ecology and Systematics*. Cornell University, Ithaca, NY.

Hill, M.O. and Gauch, H.G. (1980) Detrended correspondence analysis, an improved ordination technique. *Vegetatio*, **42**: 47–58.

Hill, M.O. (1989) Computerized matching of releves and association tables, with an application to the British National Vegetation Classification. *Vegetatio*, **83**: 187–194.

Hill, M.O. (1996) TABLEFIT version 1.0, for identification of vegetation types. *Institute of Terrestrial Ecology,* Huntingdon.

Jongman, R.H.G., ter Braak, C.J.F., and van Tongeren, O.F.R. (eds.) *Data Analysis in Community and Landscape Ecology*. Cambridge University Press, Cambridge.

Klir, G.J., and Yuan, B. (1995) *Fuzzy Sets and Fuzzy Logic: Theory and Applications.* Prentice-Hall, Englewood Cliff, NJ.

Kosko, B. (1993) *Fuzzy Thinking: The New Science of Fuzzy Logic.* Hyperion, NY.

Kruse, R., Gebhardt, J., and Klawonn, F. (1994) *Foundations of Fuzzy Systems.* Wiley & Son, Chichester.

Legendre, P. and Legendre, L. (1998) *Numerical Ecology*. Elsevier, Oxford.

Loidi, J., Herrera, M., Olano, J.M., and Silvan, F. (1994) Maquis vegetation in the eastern Cantabrian coastal fringe. *Journal of Vegetation Science*, **5**: 533–540.

McCracken, D.I. (1994) A fuzzy classification of moorland ground beetle (Coleoptera, Carabidae) and plant-communities. *Pedobiologia*, **38**: 12–27.

Moraczewski, I.R. (1993a) Fuzzy logic for phytosociology. 2. Generalizations and prediction. *Vegetatio*, **106**: 13–20.

Moraczewski, I.R. (1993b) Fuzzy logic for phytosociology. 1. Syntaxa as vague concepts. *Vegetatio*, **106**: 1–11.

Palmer, M.W. (1994) Fuzzy sets or blurry artifacts? A comment on Zhang and Oxley. *Journal of Vegetation Science*, **5**: 439–440.

Roberts, D.W. (1986) Ordination on the basis of fuzzy set theory. *Vegetatio*, **66**: 123–131.

Roberts, D.W. (1989a) Fuzzy systems vegetation theory. *Vegetatio*, **83**: 71–80.

Roberts, D.W. (1989b) Analysis of forest succession with fuzzy graph theory. *Ecological Modelling*, **45**: 261–274.

Roberts, D.W. (1996) Landscape vegetation modelling with vital attributes and fuzzy systems theory. *Ecological Modelling*, **90**: 175–184.

Rodwell, J.S. (ed.) (1991, 1992, 1995) *British Plant Communities*, volumes 1–4. Cambridge University Press, Cambridge. Published on behalf of the Joint Nature Conservation Committee with a research team of C.D. Pigott, D.A. Ratcliffe, A.J.C. Malloch, H.J.B. Birks, M.C.F. Proctor, D.W. Shimwell, J.P. Huntley, E. Radford, M.J. Wigginton, and P. Wilkins.

Smith, P.A. (1994) Autocorrelation in logistic regression modelling of species' distributions. *Global Ecology and Biogeography Letters*, **4**: 47–61.

Tsiouvaras, C.N., Nastis, A., Papachristou, T., Platis, P., and Yiakoulaki, M. (1999). Kermes oak shrubland resource availability and grazing responses by goats as influenced by stocking rate and grazing system. *In:* A. Gibon, J. Lasseur, E. Manrique, P. Masson, J. Pluvinage, R. Revilla (eds.), *Livestock Farming Systems and Land Management in the Mountain and Hill Mediterranean Regions. Options Mediterranéennes (Series B)*, **27**: 155–164.

Zhang, J.T. (1994) A combination of fuzzy set ordination with detrended correspondence analysis – One way to combine multi-environmental variables with vegetation data. *Vegetatio*, **115**: 115–121.

Zhang, J.T. and Oxley, E.R.B. (1994) A comparison of multivariate analysis of upland grasslands in North Wales. *Journal of Vegetation Science*, **5**: 71–76.

A ROUGH SET-BASED APPROACH TO HANDLING UNCERTAINTY IN GEOGRAPHIC DATA CLASSIFICATION

PIOTR JANKOWSKI,
Department of Geography, San Diego State University, 5500 Campanile Drive, San Diego, CA 92182-4493

Abstract. The chapter describes how the Rough Set-based approximation of polygon classes with point-based elementary sets can lead to classification of point-in-polygon data patterns and consequently to knowledge in terms of classification rules, which are logical statements of the "*if . . .* , *then. . .* " type. The chapter also discusses properties of Rough Set-based approximation when indiscernibility relationship is substituted with dominance relationship due to preference ordered attributes in the classification table. Since the preference order attributes are common in spatial multiple criteria evaluation the presented approach has applications in spatial decision analysis.

Keywords: Rough Sets, data classification, spatial analysis, GIS, SDSS, spatial data mining, multiple criteria evaluation

1. Introduction

Rough Set (RS) theory was introduced by Pawlak (1982, 1991) for the analysis of *inconsistent or ambiguous* description of objects. The RS philosophy is based on the assumption that every object in the universe U is associated with certain amount of information (data, knowledge). This information can be expressed by means of attributes describing the object. Objects, which have the same description are said to be indiscernible with respect to the available attributes. The *indiscernibility relation* constitutes the mathematical basis of RS theory. It induces a partition of object domain into blocks of indiscernible objects, called elementary sets, which can be used to build knowledge about real or abstract worlds. Any subset X of the universe U may be expressed in terms of blocks either precisely or approximately. In the latter case, the subset X may be characterized by two ordinary sets, called the *lower* and *upper approximations*. A rough set is

A. Morris and S. Kokhan (eds.), Geographic Uncertainty in Environmental Security, 75–87.

defined by means of these two approximations, which coincide in the case of an ordinary set. The lower approximation of X is composed of all the elementary sets whose elements certainly belong to X, while the upper approximation of X consists of all the elementary sets whose elements may belong to X. The difference between the upper and lower approximation constitutes the boundary region of the rough set, whose elements cannot be characterized with certainty as belonging or not to X using the available information. The information about objects from the boundary region is, therefore, inconsistent or ambiguous.

In the domain of spatial data handling Schneider (1997) and Worboys (1998) used the RS theory to account for imprecision resulting from spatial or semantic data resolution. The work by Ahlqvist et al. (2003) presented RS theory-based measures of uncertainty for spatial data classification. In this chapter we adopt the RS theory for a problem of multiple criteria classification where class membership is induced by both the spatial relationship of containment and attribute relationship of indiscernibility.

Consider a geographic space comprised of a set of points and a set of polygons such that each polygon contains a subset of points. Each point can be characterized by some attributes and each attribute has a defined value domain. Subsets of points indiscernible in terms of attribute values correspond to elementary sets in the sense of RS theory. We can partition the geographic space into polygons such that polygons characterized by the same property constitute a class. For example, consider a wildlife preserve partitioned into habitat areas based on the richness of species. Habitats characterized by the same richness of species constitute a class. A survey of wildlife in the preserve revealed the distribution of species along with their characteristics (attribute values). We are interested in learning whether the elementary sets of surveyed species describe the habitat classes precisely or approximately. Note that even though the description of habitat classes by surveyed species is based on attributes, it is facilitated by containment relationship resulting from point-in-polygon intersection. The lower approximation of a habitat class is composed of all species that are fully contained in the class, while the upper approximation consists of those species, which are partially contained in the class. The partial containment means that only some members of the elementary set are contained in a given polygon class while other representatives are contained in other polygon classes.

Information about point-in-polygon pattern can be stored in a point attribute table called here the *classification table* where one column represents a polygon class and the rest of columns represent point attributes.

In the classical RS theory the data about objects belonging to a set U can be either quantitative or qualitative. The classical rough set approach is, however, unable to deal with preference-ordered attribute domains and preference-ordered classes. A preference order means that either a higher attribute value is preferred to lower attribute value (benefit-based preference order) or that lower attribute value is preferred to higher attribute value (cost-based preference order. Attributes that have known preference order are called criteria in decision theory. In this paper we examine situations where domains of some attributes have an established preference order. Referring to the wildlife example, habitats can be ordered from poor, through satisfactory, to good, and species can be described by preference-ordered attributes such as abundance and fitness.

The remainder of the chapter describes how the Rough Set-based approximation of data derived from point-in-polygon containment can lead to classification of point-in-polygon data patterns and consequently to knowledge in terms of classification rules, which are logical statements of the "*if . . .* , *then. . .* " type. The chapter also discusses properties of Rough Set-based approximation when indiscernibility relationship is substituted with dominance relationship due to preference ordered attributes in the classification table. The "*if . . .* , *then. . .* " type classification rules derived from rough sets with dominance relationship are demonstrated for the example of illegal immigrants apprehended at geo-referenced point locations along the San Diego sector of the US–Mexico international border. The apprehensions are treated as points contained in specific sections of the border – represented by polygons. The rules represent classification knowledge about border crossing choices made by different groups of immigrants described by their demographic and economic conditions.

2. Rough Set Theory

The RS theory is based on the assumption that each object in the universe is associated with knowledge which can be used to classify it. The knowledge is represented in an *information system* (data table) where rows represent objects (e.g., locations) and the columns represent attributes (e.g., elevation, average temperature, type of vegetation, and distance to road). A special form of information system is called a *decision table*, where one column represents a decision (classification) and the rest of columns represent conditions (object characteristics).

More formally, an information system is a pair $A = (U, A)$, where U is a non-empty finite set of objects called the *universe*, and $A = \{a_1, . . ., a_n\}$ is a non-empty finite set of *attributes*, i.e., $a_i : U \rightarrow V_a$ for $a \in A$, where V_a is called *value set* of the attribute a_i. The decision table is then a pair $A = (U, A \cap \{d\})$, where d represents a distinguished attribute called a *decision*. In the decision table the attributes that belong to A are called *conditional attributes* or *conditions* and are assumed to be finite. The i-th decision class is a set of objects $C_i = \{o \in U: d(o) = d_i\}$, where d_i is the i-th decision value taken from decision value set $V_d = \{d_i, \ldots, d_{|V_d|}\}$.

The *indiscernibility relation* occurs when objects with the same attribute values are present in the information system. For any subset of attributes B $\subseteq$ A the *indiscernibility relation IND(*B*)* for $x, y \in U$ is defined as follows:

$$x\ IND(\mathrm{B})\ y \Leftrightarrow \forall_{a \in B} a(x) = a(y) \tag{1}$$

Indiscernible objects are not distinguishable and cannot be further classified. The indiscernibility relation thus partitions an information system into collections of indiscernible objects called *elementary sets*, which can be used to build knowledge about a real or abstract world (Slowinski et al., 2001). This fact leads to the concept of *rough set* in terms of approximation of any set *X*, where $X \subseteq U$, and the classification of elementary sets comprising the set *X* into disjoint categories. Therefore, a rough set is a pair of a *lower* and *upper* approximation of indiscernible objects, where $\underline{B}X = \{x \in U: [x]_B \subseteq X\}$ and $\underline{B}$ is the lower approximation of *X*, and $\overline{B}X = \{x \in U: [x]_B \cap X \neq 0\}$ and $\overline{B}$ is the upper approximation of *X*. The lower approximation consists of all objects which certainly belong to the set and are certainly classified as elements of that set, while upper approximation consists of all objects which possibly belong to the set and are possibly classified as elements of that set. The *boundary* or the *doubtful* region is defined as $BN_B(X) = \overline{B}X - \underline{B}X$, which is the difference between the upper and the lower approximation and is a set of elements which cannot be certainly classified as belonging to the set *X* using the set's attributes. The boundary is a non-definable set of the universe and contains objects that cannot be classified with certainty into a set.

A set *X* is an ordinary set, called an exact set if $BN_B(X) = \varnothing$, which results in $\underline{B}X = \overline{B}X$. Otherwise, if $BN_B(X) \neq \varnothing$, the set *X* is a rough set that can be approximated with some accuracy. An *accuracy of approximation* is influenced by the existence of a doubtful region where a greater doubtful region of a set yields a lower accuracy of a set. The *accuracy of approximation* is defined as follows:

$$\alpha_B(X) = \frac{|\underline{B}X|}{|\overline{B}X|},\ where\ X \neq 0 \tag{2}$$

where the quotient is represented by the lower and the upper approximation of *X*. When classifying objects, this ratio represents the percentage of possible correct decisions while the *measure of roughness* that quantifies our knowledge is obtained by the following equation: $\rho_B(X) = 1 - \alpha_B(X)$, where $0 \leq \alpha_B(X) \leq 1$ and $0 \leq \rho_B(X) \leq 1$ for any $B \subseteq A$ and $X \subseteq U$. Additionally, *quality of classification* can also be defined for individual classes as a quotient of the cardinalities of all lower approximations of the classes in which the objects set is classified and the cardinality of the object set.

Discovering dependences between attributes in an information system is a fundamental task in the RS analysis that enables reduction of unnecessary attributes. Any $B \subseteq A$ such that *IND*(A) = *IND*(B) is a *reduct* in information system *A* and *RED*(*A*) is the set of all reducts for *A*. Therefore, a reduct is the minimal subset of attributes that provides the same quality of classification as the set of all attributes. When an information table has more than one reduct the intersection of all reducts $CORE(A) = \cap\ RED(A)$ comprises the *core*. The core represents a collection of the most important attributes in the information table.

In the case of an information system that contains conditions and decisions, a reduced information table provides rules such as "*if* conditions *then* decisions". Deterministic rules are generated when objects match one or more rules indicating a unique decision; non-deterministic rules are generated when objects match one or more rules indicating multiple decisions. The number of objects satisfying the conditions of the rule determines the *strength* of the rule. The strength of the rule can be used as a measure of uncertainty of assigning objects to a decision class. Deterministic and non-deterministic rules can be generated to assign new objects to decision classes by matching rule premises with the objects characteristics.

2.1. ROUGH SET-BASED CLASSIFICATION OF DATA WITH DOMINANCE RELATIONSHIP

Searching for classification patterns with RS represents an intelligent, data-driven approach to knowledge discovery based on learning examples. The learning examples represent prior knowledge that can include ranges of attribute values, a division of attributes into condition and decision attributes and the resulting functional relationships, in which conditions describe a decision class, and a semantic correlation between pairs of attributes (Slowinski et al., 2005). Attributes that are semantically correlated have also preference-ordered domains and are called evaluation criteria since only due to an established preference order one can judge whether more is better, as in the case of benefit criteria, or less is better (the case of cost criteria). Semantically correlated criteria are an important component of decision support. In the case of sorting/classification decision problem a semantic correlation between a condition and a decision criteria means that an improvement in the value of condition should not worsen the value of decision, but rather should improve it while the values of other criteria remain unchanged. Generalizing the above statement semantic correlation between condition criteria and decision criteria requires that an object x dominating an object y on all condition criteria also dominates object y on all decision criteria. The statement that object x dominates object y means that x is at least as good as y on all evaluation criteria. This dominance principle is known as Pareto-*dominance*. Using a simple example of two houses under consideration (house x1 and house x2), location atractiveness as the decision criterion and two condition criteria including proximity to schools and price of house, if house x1 is closer to schools and costs less then house x2 then house x1 should also be more attractive than house x2.

Classification patterns of semantically correlated condition and decision criteria can be represented by "if...then..." decision rules. Each rule is represented by a condition profile and a decision profile corresponding to criteria values. A decision rule follows the dominance principle if a rule has at least one condition–decision pair of semantically correlated criteria. A rule profile dominates another rule profle if the criteria values of the former are at least as good as criteria value of the latter profile. Observing the dominance principle in discovering condition–decision classification patterns allows to uncover inconsistency in prior knowledge (learning examples). Consider the case of three houses: x1, x2, and x3. House x1 located closer to schools and costing less than house x2 and as was judged as more attractive than x2. Yet house x3 located further away from schools than house x1 and costing the same as x2 was found as attractive as house x1, which is clearly inconsistent in terms of dominance principle. In order to ensure that RS-derived decision rules be consistent with the dominance principle Greco et al. (1998) proposed an extension of RS-based knowledge discovery paradigm called the Dominance Based Rough Set Approach. The formal bases of this approach are outlined below following their presentation in Slowinski et al. (2005).

Let a pair $A = (U, A)$ denote a final and non-empty set of objects U called the *universe*, and $A = \{a_1, \ldots, a_n\}$ is a non-empty finite set of *attributes* divided into a set of condition attributes C and a set of decision attributes D such that $C \cap D = \varnothing$. Then sets X_C and X_D can be defined as attribute domains corresponding to condition and decision attributes. The elements of X_C and X_D represent possible evaluations of objects on the attributes from sets C and D, respectively. Let X_q be the set of all possible evaluations of objects with respect to attribute q, where the value of object x for attribute q is denoted as x_q. Objects x_q and y_q are indiscernible by $B \subseteq C$ if $x_q = y_q$ for all $q \in B$ and additionally, objects x_q and y_q are indiscernible by $E \subseteq D$ if $x_q = y_q$ for all $q \in E$. The sets of objects indiscernible by attributes from C and D are called the equivalence classes of the corresponding indiscernibility relation *IND*(C) or *IND*(D). The indiscernibility relation for all decision attributes *IND*(D) results in partitioning U into a finite number of decision classes $Cl = \{Cl_t, t = 1,...,n\}$ such that each $x \in U$ belongs to only one class $Cl_t \in \mathbf{Cl}$.

If pairs of condition–decision attributes are semantically correlated then the indiscernibility relation is unable to produce the equivalence classes in C or D that would represent the preference order and hence, it needs to be replaced by the dominance relation in the condition and decision attribute domains X_B and X_E ($B \subseteq C$ and $E \subseteq D$).

Let us define the dominance relation as follows. Object x dominates object y, xD_By, if $x_q \geq y_q$ for all attributes $q \in B$ and $B \subseteq C$. Additionally, one can state that object x dominates object y, xD_Ey if $x_q \geq y_q$ for all $q \in E$ and $E \subseteq D$. In the condition attribute domain X_B the dominance relations xD_By, and xD_By are directional statements where x is a subject and y is a referent. A set of objects y dominated by objects x is called a B-dominated set and it is defined as $D_B^-(x) = \{y \in U : xD_By\}$. Conversely, if x is the referent and y the subject than one can define a set of objects y dominating objects x defined as $D_B^+(x) = \{y \in U : yD_Bx\}$.

For the decision attribute domain X_E one can define the following set resulting from the E-dominance relation: $Cl_E^{\geq x}(x) = \{y \in U : yD_Ex\}$ and $Cl_E^{\leq x}(x) = \{y \in U : xD_Ey\}$. Let us define a specific decision class in D as $Clt_q = \{x \in X_D : x_q = t_q\}$. If $x \in Cl_E^{\geq x}$, then x belongs to class Clt_q or better. Conversely, if $x \in Cl_E^{\leq x}$, then x belongs to class Clt_q or worse. The equivalence classes are then defined by dominance relations $D_B^-(x)$ and $D_B^+(x)$ in the condition attribute domain and $Cl_E^{\geq x}(x)$ and $Cl_E^{\leq x}$ in the decision attribute domain. Classification patterns can then be represented as rules in which the dominance classes in the condition domain: $D_B^+(x)$ and $D_B^-(x)$ describe the dominance classes in the decision domain: $Cl_E^{\geq x}(x)$ and $Cl_E^{\leq x}$.

3. Application of Dominance-Based Rough Sets to Classification of Migrants Crossing the US–Mexico Border

The formal framework presented in section 2.1 was applied to the problem of classifying immigrants crossing illegally from Mexico to the USA along the San Diego County sector of the border (about 60 miles long). The data came from the US Border Patrol data base containing the geographical coordinates of each apprehension event between October 2003 and February 2006 and included the characteristics of apprehended individuals such as sex, age, home location, the amount paid to smugglers – if any, the number of repeated apprehension, and others. These data were then joined with another data set obtained from the 2000 Mexican Census describing economic and demographic characteristics of the origin places of apprehended immigrants. These data were collected at the level of *municipios* (the equivalent of counties in the USA).

3.1. DATA PREPARATION

The data was organized into a decision table with characteristics describing the apprehended immigrants as condition attributes. The single decision attribute used in the analysis was defined as the difficulty of crossing the border from Mexico into the USA. The *border crossing difficulty* was derived from a spatial analysis procedure involving data classification and weighted map overlay. The purpose of classification was to partition the 1-mile-wide and 60-miles-long stretch of land along the San Diego County sector of the US–Mexico border into homogeneous segments. The width of the border region was determined based on the majority of apprehensions taking place within 1-mile-wide strip of land north of the border. The classification of the border region was based on eight spatial data layers including climate, land use, proximity to major roads, enforcement effort (number of border patrol agents per border/mile), fencing (percentage of border fences per border/mile), sensors (number of sensors per border/mile), slope, and vegetation–trail index representing a combination of vegetation and trail density. The input data layers were represented in raster format in GIS and each raster cell was classified using fuzzy *K*-means classification technique. Fuzzy *k*-means clustering approach, (Burrough and McDonnell, 1998) is analogous to traditional cluster analysis. Cluster analysis or clustering is a method that groups patterns of data that in some sense belong together and have similar characteristics. The clustering technique uses a repetitive procedure by selecting a set of random cluster points and building clusters around each seed. This is accomplished by assigning every point in the data set to its closest seed, using distance measures such as: Euclidian, Mahalanobis or Diagonal distance. The iteration stops when a stable solution is reached meaning that the objects (raster cells) in each cluster are similar to one another while those in different clusters are not similar to one another.

The idea of fuzzy clustering was introduced first by Ruspini (1969) as an alternative to the traditional cluster analysis by applying membership values to points between clusters as an inverse function of distance from the cluster centers. McBratney and deGruijter (1992) refer to the fuzzy *k*-means clustering term as a "continuous classification" where each data point is not required to be an exclusive member of one and only one class. The membership value is assigned through the class centroid concept for each data point in each class. The final membership values with fuzzy *k*-means range between 0 and 1 for each data point, while the sum of values for a particular data point across all classes equals 1.

The fuzzy *k*-means classification was performed in FuzME software (Minasny and McBratney, 2000) and resulted in seven classes with each class being a unique combination of input data layers. The classes were then mapped onto the border region and the mapping became the basis for partitioning the 1-mile-wide, 60-mile-long border strip into nine segments. Each segment is characterized by a unique combination of input data layers in terms of the difficulty they pose at crossing the border. For example, segment 1 stretching from the Pacific Ocean a few miles east is characterized by the low difficulty posed by vegetation, climate, road access, slope, and land use, medium difficulty posed by sensors, and high difficulty presented by fences and enforcement effort. The overall measure of border crossing difficulty was computed for each segment with weighted overly where each of eight input data layers was assigned a weight representing its relative importance and the robustness of the result was tested by "shaking" the weights using the Monte Carlo analysis approach. The standardized scale of the border crossing difficulty was from 0 (very easy to cross) to 3 (very difficult to cross).

3.2. CLASSIFICATION RESULTS

The geographic data-set (in shapefile format) used for the classification contained the Latitude and Longitude coordinates and selected characteristics of 7,650 apprehensions that took place within the 1-mile-wide strip of the international border along the San Diego County, between October 2002 and February 2006. The 7,650 apprehension records were selected from a larger data-set based on the availability of place of birth information for each apprehended individual and its successful matching with the reference Places/Municipios database. There were 15 condition criteria and one decision criterion used in the analysis. The criteria along with their preference order (benefit – the higher the value the better, and cost – the lower the value the better) are presented in Table 1. The decision variable was represented by three classes of border crossing difficulty: easy (0–0.9), medium (0.91–1.3) and difficult (1.31–3.0).

TABLE 1. Criteria used in dominance-based rough set classification of apprehended immigrants

Criterion name	Data type	Preference order
Border crossing difficulty	Integer	Cost
Age	Continuous	Benefit
Payment to smugglers	Continuous	Benefit
Repeated apprehensions	Continuous	Benefit
Mortality rate	Continuous	Cost
Percentage of pop speaking an indigenous language	Continuous	Cost
Percentage of illiterate pop	Continuous	Cost
Percentage of employed males	Continuous	Benefit
Percentage of earnings less than min wage	Continuous	Cost
Percentage of homes with services (water, electricity, sewage)	Continuous	Benefit
Total fertility rate	Continuous	Benefit
Percentage of pop age 30–39 Living in another country in 1995	Continuous	Benefit
Businesses per males Age 20–29	Continuous	Benefit
Employees per firm	Continuous	Benefit
Wage per employee	Continuous	Benefit
Gross product per firm	Continuous	Benefit

The data were processed with 4eMKa2 software tool developed by the Laboratory of Intelligent Decision Support Systems at the Poznan University of Technology. The software is an implementation of the dominance relation-based rough set approach to multiple criteria decision support, especially for sorting/classification problems. The results comprise 1,330 rules out of which 28 rules have the strength of at least 2%, which means that they are supported by at least 153 apprehension records. These rules classify the immigrant profiles, based on the values of 15 condition criteria, into the "medium" class of border crossing difficulty. Two characteristics common to all 28 rules are the age of apprehended immigrants equal or less than 24 years (sample mean = 28.8) and total fertility rate equal or less than 3 (sample mean = 2.98). Three strongest rules are listed below. The condition criteria values are accompanied by the sample means given in parenthesis.

Rule #1 (relative strength = 2.42%): IF Age $\leq$ 24 (28.8) AND Morality $\geq$ 4.52% (5.29%) AND Percentage of employees earning less then minimum wage ≥ 10.1% (10.1%) AND Total fertility rate ≤ 3 (2.98) AND Percentage of population living abroad ≤ 0.5% (1%) THEN Difficulty ≤ Medium.

Rule #2 (relative strength = 2.4%): IF Age ≤ 23 (28.8) AND Percentage of employees earning less then minimum wage ≥ 19.8% (10.1%) AND Percentage of houses with basic service ≤ 51.1% (64.4%) AND Total fertility rate ≤ 3 (2.98) AND Number of employees per firm ≤ 5.42 (5.19) THEN Difficulty ≤ Medium.

Rule #3 (relative strength = 2.37%): IF Age ≤ 23 (28.8) AND Percentage of houses with basic service ≤ 52.8% (64.4%) AND Total fertility rate ≤ 3 (2.98) AND Number of employees per firm ≤ 4.48 (5.19) AND Wage per employee ≤ $3.1 ($3.4) THEN Difficulty ≤ Medium.

These and the other 25 rules represent the immigrant profiles that are similar to the mean values of condition criteria with some bias towards the below-mean values (e.g., age, below minimum wage earners, houses with basic services).

In summary, no rules with relative strength of 10% or more were found. About 2% (28) of all 1,330 rules had the relative strength of 2% or more. The strongest rule had the relative strength of 2.42% and the support of 176 out of 7,650 records. However, 2% of all the rules classified 52.7% (4,039) of all of the apprehension records. Common condition profiles represented by these rules were close to the mean values characterizing the economic conditions of the sample with some bias towards younger age of immigrants and worse than average economic conditions. These condition profiles were classified in the medium class of border crossing difficulty.

4. Conclusion

The purpose of classifying data about illegal immigrants apprehended in the San Diego County sector of the US–Mexico border was to test whether or not a knowledge discovery approach based on rough sets would result in strong predictive rules. Such rules could theoretically link various common immigrant profiles, given by the values of condition criteria, with the classes of border crossing difficulty thus revealing possible border crossing preferences based on demographic and economic conditions. This reasoning is based on simple, yet unconfirmed theoretical arguments that would-be immigrants coming from better economic conditions have more resources at their disposal to choose border sectors that are less difficult to cross then those who come from poorer economic conditions. Similarly one can speculate that those with some experience abroad (usually in the USA) and previous apprehension experience will be more savvy in selecting their point of entry into the USA than those who lack such experience. These theoretical arguments were used in defining the preference order for condition criteria used in the classification. The lack of stronger and more discriminating rules than those derived with dominance-based rough set classification may result from three possible causes. First, theoretical assumptions behind the preference order of condition criteria may be wrong. Second, the quality of prior knowledge given by the condition criteria may be inadequate and it may require different conditions describing both socio-economic characteristics of immigrants and social networks facilitating their journeys to the USA. Third, alternative partitions of the decision criterion domain can be used as decision classes. One can also use different decision variables based on alternative definitions of border crossing difficulty.

Acknowledgement

This research was supported by the NASA-funded REASoN grant "A Border Security Decision Support System Driven by Remotely Sensed Data Inputs." Any opinions, findings, and conclusions or recommendations expressed in this material are those of the author and do not necessarily reflect the views of NASA.

Bibliography

Ahlqvist, O., Keukelaar, K., and Oukbir K., 2003, Rough and fuzzy geographical data integration, *International Journal of Geographical Information Science*, **17**(3): 223–234.

Burrough, P.A., and McDonnell, R.A., 1998, *Principles of Geographic Information Systems.* Oxford University Press, Oxford.

Greco, S., Matarazzo, B. and Slowinski, R., 1998, A new rough set approach to evaluation of bankruptcy risk, in: Zopounidis, C. (ed.), *Operational Tools in the Management of Financial Risk*, Kluwer, Dordrecht, pp. 121–136.

McBratney, A.B., deGruijter, J.J., 1992, A continuum approach to soil classification by modified fuzzy k-means with extragrades, *Journal of Soil Science*, **43**: 159–175.

Minasny, B., and McBratney A.B., 2000, FuzME version 2.1, Australian Centre for Precision Agriculture, The University of Sydney, Accessed 18 November 2005, http://www.usyd.edu.au/su/agric/acpa).

Pawlak, Z., 1982, Rough sets, *International Journal of Computer Information Sciences*, **11**(5): 341–356.

Pawlak, Z., 1991, *Rough Sets: Theoretical Aspects of Reasoning About Data.* Kluwer, Dodrecht.

Ruspini, E.H., 1969, A new approach to clustering, *Information Control*, **15**: 22–32.

Schneider, M., 1997, *Spatial Data T types for Database Systems*, Lecture notes in computer science, Vol. 1288, Springer, Berlin.

Slowinski, R., Greco, S., and Matarazzo B., 2005, Rough set based decision support, in: E.K. Burke and G. Kendall (eds.), *Search Methodologies: Introductory Tutorials in Optimization and Decision Support Techniques*, Springer, NY, pp. 721–778.

Worboys, M.F., 1998, Computation with imprecise geospatial data, *Computers, Environment and Urban Systems*, **22**: 85–106.

FUZZY MODELS FOR HANDLING UNCERTAINTY IN THE INTEGRATION OF HIGH RESOLUTION REMOTELY SENSED DATA AND GIS

JOCHEN SCHIEWE*, MANFRED EHLERS
University of Osnabrueck, Institute for Geoinformatics and Remote Sensing, Seminarstr. 19a/b, 49069 Osnabrueck, Germany

Abstract. The advent of new high resolution sensors, either airborne or spaceborne, leads to new applications and further impulses for an integration of remotely sensed and GIS data. Along with these new data sources, existing processing methods have to be adopted which is in particular also valid for the assessment of the post classification quality. In this overall context our contribution will outline general uncertainty aspects in the integration process and particular problems with the accuracy assessment based on high resolution data. These problems lead to the motivation to develop a new characteristic value, called the Fuzzy Certainty Measure (FCM), which considers indeterminate boundaries in the classification result as well as in the reference data, and can be applied in a class- and even object-specific manner.

Keywords: classification, fuzzy, high resolution, indeterminate boundaries, integration, object-based, remote sensing, uncertainty

1. Introduction

The development and usage of new digital remote sensing system – either spaceborne (like Ikonos, OrbView or QuickBird), or airborne (like HRSC, ADS 40, DMC) – has answered user demands concerning improved spatial, spectral, and radiometric resolutions. Along with that a variety of new thematic applications, in general for generating and updating GIS databases at large scales, becomes possible. In return, GIS information can be used to

*Jochen Schiewe, University of Osnabrueck, Institute for Geoinformatics and Remote Sensing, Seminarstr. 19a/b, 49069 Osnabrueck, Germany, E-Mail: jschiewe@igf.uni-osnabrueck.de.

A. Morris and S. Kokhan (eds.), Geographic Uncertainty in Environmental Security, 89–106.

support and automate the object extraction process. However, the current operational status of the *integration of GIS and remotely sensed data* can be described as not satisfying yet as section 2 will show.

With the new high resolution data also new techniques for their interpretation have become necessary. In the meantime region-based methods, i.e., segmentations, as well as follow-up object-based and eventually fuzzy classification approaches, are widely applied. However, an integrated *quality assessment* of the interpreted remotely sensed and the GIS data is still a general problem because neither the transformation nor the propagation of the respective measures is straightforward. Furthermore, additional problems arise with the use of high resolution data when standard methods for determining the post classification quality are applied. Section 3 discusses both, general uncertainty aspects in the integration process, and particular problems with the post classification accuracy assessment based on high resolution data. In this context some deficits are elaborated which lead to the motivation to develop a new characteristic value, called the *Fuzzy Certainty Measure (FCM)*, which considers indeterminate boundaries in the classification result as well as in the reference data. Section 4 explains the underlying concept and possible extensions of this measure.

2. Integration of Remote Sensing and GIS

The ideal goal of integration should be that GIS objects can be extracted from a remote sensing image to update the GIS database. In return, GIS "intelligence" (e.g., object and analysis models) should be used to support and automate this object extraction process. In this context as early as 1989, Ehlers et al. (1989) presented a concept for a totally integrated system for remote sensing and GIS. They differentiated between three integration levels: (a) two separate systems with a data interface; (b) two principally separate systems with a common user interface; and (c) a totally integrated system.

Most of today's GISs offer hybrid processing, that is, the analysis of raster and vector data. They also have image display capabilities or image analysis add-ons which offer some level (b) functionality (Bill, 1999; Ehlers, 2000).

However, geospatial information is usually processed in either raster or vector form and has to be converted into the desired processing or output format. A truly integrated processing option (without prior conversion) does not really exist. This is also valid for integrated remote sensing/GIS analyses. Efforts in this direction have usually been restricted to a simple geometric registration and overlay process. An integrated system of image and GIS data is still missing and GIS and remote sensing information is usually processed independently from each other.

The requirements for integrated systems are usually defined on an ad hoc basis which is driven by project demands or the data sources to be incorporated. What is needed for a general concept is an analysis of the necessary processing components of such an integrated system. The data integration approach has to be replaced by an analysis integration approach. This implies that we need to identify a set of analytical geoinformatics functions that are software system independent.

If one looks into the functionality of current GIS it is immediately evident that GIS operations are usually based on the underlying system and its associated data structure. A general description of GIS functions could offer a system independent view. First attempts could be seen by Tomlin's Cartographic Modeling (Map Algebra) approach which is still the basis for many raster GIS's (Tomlin, 1990). Another approach for raster and vector based systems was presented by Albrecht who proposed a set of system and data structure independent GIS operators (Albrecht, 1996).

Also in remote sensing we experience very diverse approaches towards image analysis taxonomies. Even a cursory look at textbooks shows that often hardware, sensors, systems, and operations or mixed together or present structures that are inconsistent with a rational image processing taxonomy. This inconsistency when dealing with image analysis taxonomies is an impediment for the development of a stronger theoretical background for the design and implementation of integrated GIS. Without such a theoretical basis, however, the only way to a GIS/remote sensing integration seems to be a project driven ad hoc approach with limited usefulness and applicability.

To define a taxonomy of data structure and system independent GIS/image analysis functions one has to start either from the remote sensing or the GIS side. Based on typical remote sensing analyses, Ehlers (2000) proposed four groups with 17 image processing functions to be added to the 20 universal GIS operators. It has to be noted however, that these operators are not sufficient to define and describe the complete functionality of integrated GIS. Still required is a thorough analysis of hybrid processing capabilities, that is, functions that allow a joint analysis of remote sensing and GIS information. It has to be investigated how polymorphic techniques can be used to extend the capacities of the universal high level GIS/image processing functions. The operator *Overlay*, for example, should be able to process image-image, GIS-image, and GIS-GIS overlays without a different name for every function option. First results of such polymorphisms were investigated, for example, by Jung (2004). Additional functions have to be developed, on the other hand, that extend the capabilities of integrated GIS beyond the sum of the single components. Three-dimensional urban information systems created from GIS and remote sensing can be seen as an example for these extensions.

It is also obvious that such a framework has to cover the topic of an integrated error assessment if the advantages of integrated processing can indeed be realized. In fact, the operator "Error Assessment" appears twice by Ehlers (2000), once for geometric operators and once for feature extraction.

3. Uncertainty Determination for High Resolution Remotely Sensed Data

3.1. GENERAL UNCERTAINTY ASPECTS

The advantages of an integrated geoprocessing framework have been proven by many examples. It is however, also evident that the issue of accuracy and errors within this integration process have to be addressed. Good science requires statements of accuracy, by which the reliability of results can be understood and communicated. Where accuracy is known objectively then it can be expressed as *error*, where it is not, the term *uncertainty* applies (Hunter and Goodchild, 1993). Thus uncertainty covers a broader range of doubt or inconsistency, and includes error as a component. The understanding of uncertainty as it exists in geographic data remains a problem that is only partly solved (e.g., Story and Congleton, 1986; Goodchild and Gopal, 1989; Veregin, 1995; Ruiz, 1997; Worboys, 1998; Gahegan and Ehlers, 2000; Zhang and Goodchild, 2002). However, without quantification, the reliability of any results produced remains problematic to assess and difficult to communicate to the user. GISs provide a whole series of tools with which data can be manipulated, without offering any control over misuse.

Uncertainty in its many forms has been on the research agenda of the GIS and remote sensing community for at least two decades, gaining much of its early momentum from the very first research initiative of the US National Center for Geographic Information and Analysis (NCGIA) (Goodchild and Gopal, 1989). Work to-date on uncertainty addresses the inherent errors present within specific types of data structure (e.g., raster or vector) or data models (e.g., field or object). The affects of error propagation and analysis within these various paradigms have been studied by Veregin (1995), Openshaw et al. (1991), Goodchild et al. (1992), Heuvelink and Burrough (1993), Ehlers and Shi (1997), Leung and Yan (1998), Shi (1998), Arbia et al. (1999), Zhang and Kirby (2000), Zhang and Stuart (2001), Shi et al. (2003). In a recent compendium on uncertainty in geographic information, Zhang and Goodchild (2002) investigate methods for uncertainty assessment for continuous variables (fields), categorical variables (classes), and objects. Despite the progress made to-date, they conclude that "academics, technologists, government information agencies, the general public, and the commercial sector must work together to take advantage of the benefits of geographical information in new applications, while being fully informed of the nature and implications of the associated uncertainties. Scientists and workers lead the leap forward".

Even if we restrict uncertainty description to one specific problem, the integration of remote sensing and GIS, a generally applicable solution does not exist. Remote sensing scientists have always had the need to quantify errors that were associated with the processing of remotely sensed data. Most efforts have gone into the error analysis of rectification and registration processes and of information extraction or multispectral classification techniques (e.g., Ehlers, 1997; Congalton and Green, 1999).

Gahegan and Ehlers (2000) formulated a framework for modelling uncertainty in an integrated remote sensing environment. A typical path taken by data captured by satellite, then abstracted into a suitable form for GIS is shown in Figure 1, and involves four such models. Continuously varying fields are quantized by the remote sensing device into image form, then classified and finally transformed into discrete mapping objects. The overall object extraction process is sometimes referred to as semantic abstraction due to the increasing semantic content of the data as it is manipulated into forms that are easier for people to work with.

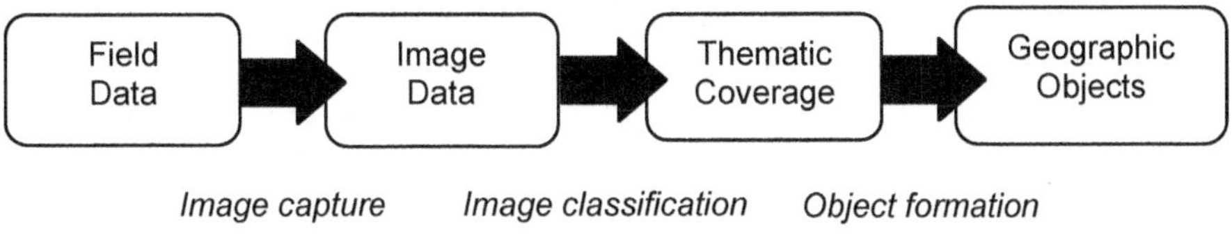

Figure 1. Continuum of abstraction from field model to object model (after Gahegan and Ehlers, 2000)

When transforming data between different conceptual models of geographic space, the uncertainty characteristics in the data may change, because techniques used to transform the data also alter the inherent uncertainty and in addition may introduce further uncertainty of their own. The four stages shown in Figure 1 represent the four models of geographic space, namely field, image, thematic, and object (or feature) models and are typical of those used in the integration of GIS and remote sensing activities. These models represent the conceptual properties of the data only and can be considered here as independent from any particular data structure that might be used to encode and organize the data.

Based on this model, Gahegan and Ehlers (2000) developed an integrated error simulation model for the transition from field (raw remote sensing) data to geo-objects. The description of uncertainty followed that proposed by Sinton (1978). It covers the sources of error as they occur in remote sensing and GIS integration. Uncertainty is restricted to the following properties:

Value (including measurement and label errors)
Spatial
Temporal
Consistency,
Completeness

Of these, measurement and label errors as well as uncertainties in space and time can apply either individually to a single datum or to any set of data. The latter two properties of consistency and completeness can only apply to a defined data-set since they are comparative (either internally amongst data or to some external framework). Table 1 summarizes this concept.

New research on uncertainty deals with the development of advanced processing techniques for the information extraction from remotely sensed images. The inclusion of contextual information (textures, neighbourhood), object or segment-based analysis techniques together with the application of fuzzy set theory and artificial intelligence challenge the standard image processing strategies (Wang, 1993; Ryherd and Woodcock, 1996; Lucieer and Stein, 2002; Ibrahim et al., 2005).

TABLE 1. Types of uncertainty and their sources in four models of geographic space (from Gahegan and Ehlers, 2000)

Type of uncertainty	Models of geographic space			
	Field	Image	Thematic	Object
Value	Measurement error and precision	Quantization of value in terms of spectral bands and dynamic range	Labelling uncertainty (classification error)	Identity error (incorrect assignment of object type), object definition uncertainty
Space	Locational error and precision	Registration error, sampling precision	Combination effects when data represented by different spatial properties are combined	Object shape error, topological inconsistency, "split and merge" errors
Time	Temporal error and precision	(Temporal error and precision are usually negligible for image data)	Combination effects when data represent-ing different times is combined	Combination effects when data representing different times is combined
Consistency	Samples/ readings collected or measured in an identical manner	Image captured identically for each pixel, but: inconsistencies due to medium between satellite and ground sensing, light fall-off, shadows	Classifier strategies are usually consistent in their treatment of a data-set	Methods for object formation may be consistent, but often are not; depends on extraction strategy
Completeness	Sampling strategy covers space, time, and attribute domains adequately	Image is complete, but parts of ground may be obscured (clouds, trees)	Completeness depends on the classification strategy (classifying entire data-set or only some classes?)	Depends on extraction strategy; spatial and topological inconsistencies may arise as a result of object formation

3.2. POST CLASSIFICATION ASSESSMENT FOR HIGH RESOLUTION DATA

In the following we will focus on one element of the "uncertainty chain" as presented in Figure 1, namely the thematic uncertainty which arises after the classification of remotely sensed scenes. The corresponding evaluation, which determines the quality of the input data and the classification process as such, seems to be a standard task: Quantitative methods compare reference data ("ground truth") and the classification result from which error matrices and related measures like overall, producer's and user's accuracy, or Kappa coefficient can be derived.

However, in the case of using spatial high resolution data, some of the general problems related to this procedure are even amplified and need even more attention compared with the use of lower resolution data. The underlying reasons, which will be briefly discussed in the following, can be grouped into geometric and semantic aspects.

From a *geometric point of view* the smaller pixel sizes lead to the fact that a suitable reference with appropriate positional accuracy as well as little model and cartographic generalization is more difficult to find.

Furthermore, an adoption of the number and size of sample units has to take place. In particular the conventional acquisition on per pixel basis is not suitable anymore due to too small elements and neglecting the neighbourhood. In analogy to the object-based interpretation approach, a per-object sampling seems to be necessary in order to define training and test elements. However, due to missing methodology such an approach is hardly applied in practice.

It is also well known that we have to handle indeterminate boundaries or spatial transition zones between mostly natural objects (e.g., between forest and meadow), which are in some cases also a function of time (e.g., the boundary between beach and water). On the other hand, we have also to consider blurred or overlapping definitions of classes or related attributes in a given classification scheme. Taking now high resolution data into account, the absolute number of pixels describing spatial transition zones – and with that the effect of fuzziness – increase.

From a *semantic point of view*, high resolution data allow for the extraction of more thematic details and object classes. With that a more complex classification scheme becomes necessary which on the other hand inherits a greater chance of overlapping definitions of attribute value ranges. As a consequence, this may lead to errors or ambiguous assignments during the visual or automatic interpretation process. The greater number of possible classes also makes more sampling units necessary. Like with geometric properties it is also difficult to find a suitable reference with appropriate thematic details and semantic accuracy. It has to be kept in mind that very often a reference data-set is nothing else than another classification result based on another, eventually lower resolution data-set.

Finally, the spatial variance within regions representing a topographical object is increased which leads to more objects and with that to more mixed objects (e.g., forest consists of trees, bare soil, and others) as well as to more boundaries. With the latter the number of indeterminate boundaries, in other words the effect of fuzziness, is again increased.

Combined together all these effects lead into classification results as demonstrated in Figure 2 where data of the digital airborne camera system HRSC-AX that delivers image and elevation data in very high spatial resolution (here: ground pixel size equals to 15 cm) has been processed. In conclusion, there is a significant necessity to develop uncertainty measures that consider uncertainties in reference data as well as indeterminate boundaries in both, reference data and classification results.

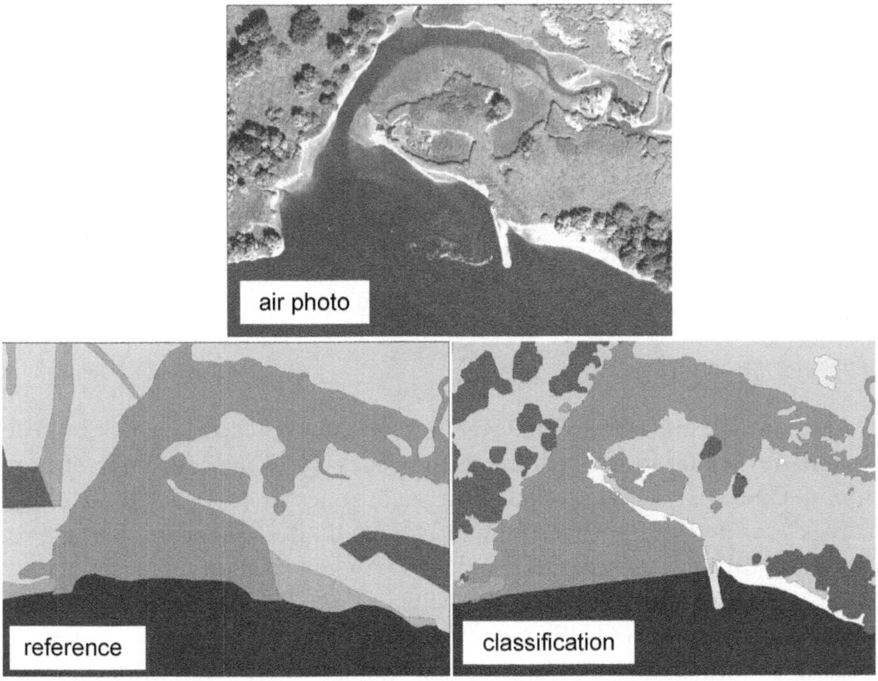

Figure 2. Top: Given digital air photo (HRSC-AX, 15 cm ground pixel size), bottom left: reference from field survey, bottom right: image classification result (Ehlers et al., 2006)

4. Alternative Approach

4.1. GOAL

Motivated by the above outlined problems, our goal is to develop a profound fuzzy methodology for determining the classification accuracy for high resolution data. In this context it has to be stated that while for the application of conventional, statistically founded methods a variety of papers exist (e.g., Thomas et al., 2003; Foody, 2004), fuzzy approaches have been considered rather rarely. One significant influence came from Gopal and Woodcock (1994) who added certainty values on a linguistic scale ("absolutely safe", etc.) to their visually classified elements. Those linguistic values can be combined using fuzzy logic theory for a better understanding of the resulting map. Similar approaches are reported by Wang and Hall (1996), Townsend (2000), or Lowell et al. (2005).

In our case we have to determine the classification accuracy on a metric scale considering fuzzy boundaries in both, the classification result and the reference data-set. We assume that an appropriate sampling procedure has been taken into account, the classification schemes between reference and classification are identical, and no discrepancies occur due to different pixel sizes or temporal changes. After describing the process of fuzzification, that is, the definition of transition zones and membership functions (section 4.2.), alternative characteristic values will be defined that describe the correctly and incorrectly classified elements (section 4.3.). Finally, the extension of these measures towards object-specific values is outlined (section 4.4.).

4.2. FUZZIFICATION

In section 3.2. we outlined the reasons for the existence of indeterminate boundaries which in return leads to the necessity of introducing *transition zones*. One approach for modelling such zones is to use the so called ε–bands, as defined by Blakemore (1994; cited after Ehlers and Shi, 1997). Here, the different chances of a point-in-polygon relation are described by five *qualitative* measures ("definitively in", etc.). Ehlers and Shi (1997) propose to use a probabilistic model in order to give a *quantitative* description which also allows for the combination with values of thematic uncertainty: With the application of the S-band model positional and thematic uncertainty values are linked by using the product rule. Other options to treat indeterminate boundaries (e.g., least squares polynomial fitting, corridor techniques, etc.) are listed by Edwards and Lowell (1996).

In order to model the inherent indeterminate boundaries between different topographical objects, we now introduce *fuzzy transition zones.* Their design is mainly influenced by the thematic combination of objects (e.g., the transition zone between forest and meadow is obviously larger than those between building and road). While Edwards and Lowell (1996) define the width of these zones for all pairs of object classes ("twains") from the mean deviations derived from multiple digitizations in aerial images, we favour a more practical way, namely the user-defined generation of a look-up matrix. It is obvious that this process depends on the specific application and needs expert knowledge (Congalton and Green, 1999).

Edwards and Lowell (1996) also found that not only the thematic class memberships but also the size and shape of the object areas under consideration have a significant influence on the width of the transition zone. Respective information can be combined numerically with the look-up matrix values by additional factors. For example, based on the hypothesis that the smaller an object area is, the higher the uncertainty for this object becomes, a scaling factor ranging from of 1-δ to 1+δ can be introduced as follows:

$$area\_factor = \left\{ \begin{array}{ll} \frac{-\delta}{A_{mean}} A + (1+\delta) & for\ A < 2 \cdot A_{mean} \\ 1-\delta & else \end{array} \right\} \tag{1}$$

where:

A: area size of current object under investigation

A_{mean}: average area size in entire scene

δ: interval for scaling factor

Within the now-defined transition zone and perpendicular to the boundary a *membership function* (in our case presently a linear function) is applied for each class c, which results in membership values $\mu(c)$ (see Figure 3). This procedure is performed for both, the classification result (leading to membership values $\mu_{CLASS}(c)$) and the reference data ($\mu_{REF}(c)$).

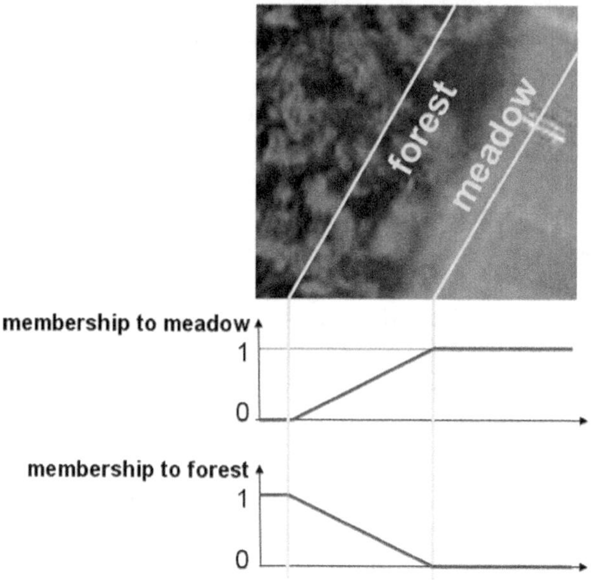

Figure 3. Definition of transition zone from look-up table, and assignment of respective fuzzy membership function

4.3. CHARACTERISTIC VALUES

Based on fuzzy membership values the goal now is to determine characteristic values that give an indication of correctly and incorrectly classified elements. In literature and system implementations only rather general measures based on fuzzy memberships are given (if at all), e.g.,

Vagueness (difference of total classification membership from maximal value 1) and

Ambiguity (difference between maximum and second largest membership value for a pixel or region)

Furthermore, only little efforts have been made for designing fuzzy confusion matrices, exemplary attempts have been undertaken by Woodcock and Gopal (2000) or Congalton and Green (1999) who used plus and minus class tolerances which make sense for continuous or ordinal, but not for nominal variables like in the case of land use classification.

Hence, our approach for determining how accurately a given class in the reference is detected in the classification result consults the respective membership values for the same spatial elements (i.e., pixels or regions).

Those are compared separately for each topographical class and for those elements for which a possibility of existence (or membership values larger than 0, respectively) is present in reference and classification. The resulting *FCM(c) per class c* is determined as follows:

$$FCM(c) = 1 - \frac{1}{n}\sum_{i=1}^{n} | \mu_{i,REF}(c) - \mu_{i,CLASS}(c) | \qquad (2)$$
$$\forall i \mid \mu_{i,REF} > 0 \wedge \mu_{i,CLASS} > 0$$

where:
$\mu_{REF}(c)$: membership value of a pixel or region for class c in reference data
$\mu_{CLASS}(c)$: *m*embership value of a pixel or region for class c in classification result
n: number of pixels or regions under consideration

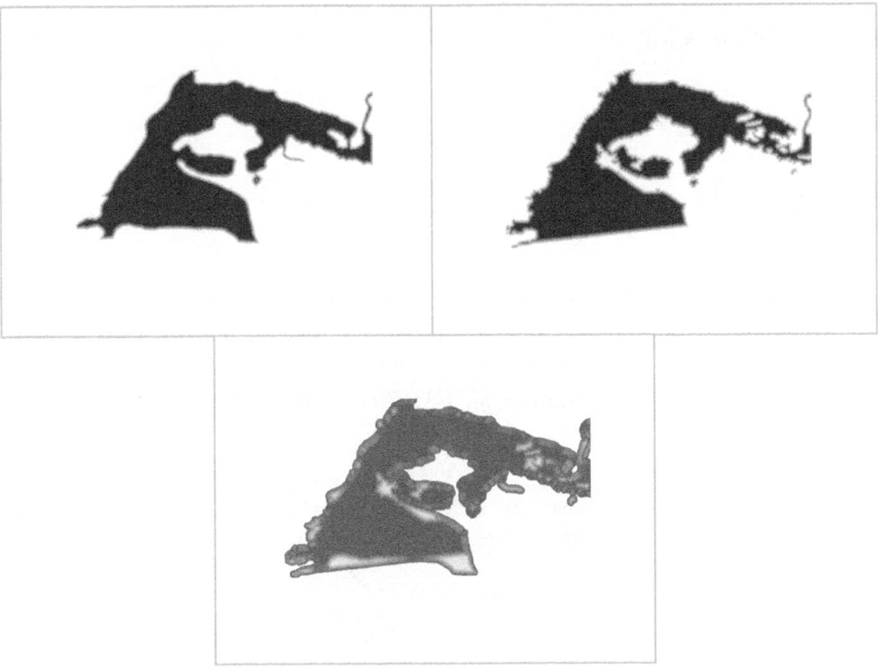

Figure 4. Top: Visualization of membership values for class KPS (tidal creek/tideland) in reference (left) and classification result (right), bottom: derived FMC values (compare to Figure 2). In all cases: The darker the area, the higher the respective value (white = 0)

The $FCM(c)$ values vary between 0 and 1 – the larger the coincidence between reference and classification is, the larger the $FCM(c)$ value becomes. Figure 4 visualizes the derivation of the FCM for a selected class within our test data-set (compare to Figure 2).

For a more thorough description of the classification uncertainty also the confusion between different classes (here named A and B) is of interest. When introducing fuzzy membership values, the major difference to a conventional confusion matrix based on crisp boundaries is that the membership areas (i.e., those regions for which the membership values are larger than 0) can overlap each other so that their class-specific values generally sum up to a value larger than 100%. And while for the computation of off-diagonal elements for conventional confusion matrices the number of misclassified elements is counted, in the fuzzy case it makes no sense just to compare the membership values for class A in the reference and class B in the classification results because the resulting difference might be correct in reality. Instead of that we consider the change in ambiguity between classes A and B in the reference compared to that in the classification as follows:

$$FMC(c_{AB}) = \frac{1}{2}\sum_{i=1}^{n} \left| \left[\left(\mu_{i,REF}(c_A) - \mu_{i,REF}(c_B) \right) - \left(\mu_{i,CLASS}(c_A) - \mu_{i,CLASS}(c_B) \right) \right] \right| \qquad (3)$$

Also for this confusion measure (with values ranging from 0 to 1) it holds that the larger the confusion is, the larger the $FCM(c_{AB})$ value becomes. Table 2 summarizes the certainty measures for the given example (diagonal elements represent $FMC(c)$ values, off-diagonal elements $FCM(c_{AB})$ values).

TABLE 2. Fuzzy Certainty Measures (FCM) for object classes under consideration (class names according to specific object catalogue, Ehlers et al., 2006)

	Classes		Reference				
			KPS	BAT	FWR	FZT	WWT
Classification	Tidal creek/Tideland	KPS	0.90	0.05	0.09	0.05	0.06
	Shrubs (Salix)	BAT		0.80	0.12	0.04	0.30
	Reed (Phragmites)	FWR			0.81	0.07	0.23
	Tidal River	FZT				0.93	0.05
	Willow Forest (Salix)	WWT					0.66

4.4. OBJECT-SPECIFIC FCM

So far, with *FCM(c)* and $FCM(c_{AB})$, respectively, the certainty of all elements belonging to a certain class in a given scene is computed at once. However, from this the uncertainty of individual objects as well, variations or outliers among different objects of the same class cannot be evaluated. Hence, an *object-specific* value is desired. There exist already some definitions by Zhan et al. (2005) for per-object measures (like: thematic, location, or shape similarity) based on *crisp* boundaries. On the other hand it is no problem to transfer the above introduced FCM to an object-specific FCM (OFCM). If the data model does not support a distinction of such individual objects, the necessary separation is performed either by introducing α-cuts or by applying a spatial segmentation (i.e., summing up membership values μ only as long as there is an N8-neighbour with $\mu > 0$). We have not applied these measures to our test data-set yet due to limited sample sizes.

5. Summary and Conclusions

High resolution remotely sensed data with ground pixel sizes of 1 m or less are a valuable data source for generating or updating large-scale GIS databases. In this context we addressed the specific problems that arise when determining the uncertainty after a classification based on high resolution data. In particular, indeterminate boundaries have to be taken into account in both, reference data and classification results.

In order to consider these fuzzy effects, we propose the introduction of a new characteristic value, the *Fuzzy Certainty Measure* (*FCM*) which is based on a posteriori definition of transition zones along object boundaries. The procedure can be characterized as flexible and as quite simple to apply. Furthermore, several extensions – in particular towards an object-specific *FCM* instead of the so far described class-specific value – can be applied without problems.

Our future work is concerned with a sensitivity analysis of the parameters (in particular with the width of transition zones based on object class combination and area sizes). Furthermore, empirical investigations will be performed for the combination of fuzzy with additional probabilistic measures. Finally, also the extension towards a change analysis can be taken into consideration by introducing thresholds for *FCM* values for the classifications of different time stamps.

References

Albrecht, J., 1996, Universal Analytical GIS Operations. Ph.D. Thesis, ISPA-Mitteilungen 23, University of Vechta, Germany.

Arbia, G., Griffith, D., and Haining, R., 1999, Error Propagation Modelling in Raster GIS: Adding and Rationing Operations. Cartography and Geographic Information Science, **26**: 297–315.

Bill, R., 1999, GIS-Produkte am Markt – Stand und Entwicklungstendenzen. Zeitschrift für Vermessungswesen, **6**: 195–199.

Congalton, R. and Green, K., 1999, Assessing the Accuracy of Remotely Sensed Data: Principles and Practices, CRC/Lewis Press, Boca Raton, FL, 137 p.

Edwards, G. and Lowell, K.E., 1996, Modeling Uncertainty in Photointerpreted Boundaries. Photogrammetric Engineering & Remote Sensing, **62**(4): 337–391.

Ehlers, M., 1997, Rectification and Registration. In: Star, J.L., J.E. Estes, and K.C. McGwire (eds.). Integration of Remote Sensing and GIS, Cambridge University Press, New York, NY, pp. 13–36.

Ehlers, M., 2000, Integrated GIS – From Data Integration to Integrated Analysis. Surveying World, **9**: 30–33.

Ehlers, M., Edwards, G., and Bédard, Y., 1989, Integration of Remote Sensing with GIS: A Necessary Evolution. Photogrammetric Engineering and Remote Sensing, **55**: 1619–1627.

Ehlers, M. and Shi, W., 1997, Error Modelling for Integrated GIS. Cartographica, **33**(1): 11–21.

Ehlers, M., Gähler, M., and Janowsky, R., 2006, Automated Techniques for Environmental Monitoring and Change Analyses for Ultra High-resolution Remote Sensing Data. Photogrammetric Engineering and Remote Sensing, **72**(7): 835–844.

Foody, G.M., 2004, Thematic Map Comparison: Evaluating the Statistical Significance of Differences in Classification Accuracy. Photogrammetric Engineering and Remote Sensing, **70**(5): 627–633.

Gahegan, M. and M. Ehlers, 2000. A Framework for Modeling of Uncertainty in an Integrated Geographic Information System. ISPRS Journal of Photogrammetry and Remote Sensing, **55**: 176–188.

Goodchild, M.F. and Gopal, S. (eds.), 1989, The Accuracy of Spatial Databases, Taylor & Francis, London.

Goodchild, M.F., Guoqing, S., and Shiren, Y., 1992, Development and Test of an Error Model for Categorical Data. International Journal of Geographical Information Systems, **6**: 87–104.

Gopal, S. and Woodcock, C., 1994: Theory and methods for Accuracy Assessment of Thematic Maps Using Fuzzy Sets. Photogrammetric Engineering and Remote Sensing, **60**(2): 181–188.

Heuvelink G.B.M. and Burrough, P.A., 1993, Error Propagation in Cartographic Modelling Using Boolean Logic and Continuous Classification. International Journal of Geographical Information Systems, **7**: 231–246.

Hunter, G.J. and Goodchild, M.F., 1993, Mapping Uncertainty in Spatial Databases: Putting Theory into Practise. Journal of Urban and Regional Information Systems Association, **5**: 55–62.

Ibrahim, M.A., Arora′, M.K., and Ghosh, S.K., 2005, Estimating and Accommodating Uncertainty Through the Soft Classification of Remote Sensing Data. International Journal of Remote Sensing, **26**: 2995–3007.

Jung, S., 2004, HYBRIS: Hybride räumliche Analyse Methoden als Grundlage für ein integriertes GIS, Ph.D. Thesis, University of Vechta, Germany (CD Publication).

Leung, Y. and Yan, J., 1998, A Locational Error Model for Spatial Features. International Journal of Geographical Information Science, **12**: 607–620.

Lowell, K., Richards, G., Woodgate, P., Jones, S., and Buxton, L., 2005, Fuzzy Reliability Assessment of Multi-Period Land-cover Change Maps. Photogrammetric Engineering and Remote Sensing, **71**(8): 939–945.

Lucieer, A. and Stein, A., 2002, Existencial Uncertainty of Spatial Objects Segmented from Satellite Sensor Imagery. IEEE Transactions on Geoscience and Remote Sensing, **40**: 2518–2521.

Openshaw, S., Charlton, M., and Carver, S., 1991, Error Propagation: a Monte Carlo Simulation. In: Masser, I., and M. Blakemore (eds.). Handling Geographic Information, Longman, London, pp. 78–101.

Ruiz, M.O, 1997, A Causal Analysis of Viewshed Error. Transactions in GIS, **2**: 85–94.

Ryherd, S. and Woodcock, C., 1996, Combining Spectral and Textural Data in the Segmentation of Remotely Sensed Images. Photogrammetric Engineering and Remote Sensing, **62**: 181–194.

Shi, W.Z., 1998, A Generic Statistical Approach for Modelling Errors of Geometric Features in GIS. International Journal of Geographical Information Science, **12**: 131–143.

Shi, W.Z., Cheung, C.K., and Zhu, C.Q., 2003, Modelling Error Propagation in Vector-Based GIS. International Journal of Geographical Information Science, **17**: 251–271.

Sinton, D., 1978, The Inherent Structure of Information as a Constraint to Analysis: Mapped Thematic Data as a Case Study. In: Dutton, G. (ed.), Harvard Papers on Geographic Information Systems, 6, Addison Wesley, Reading, MA.

Story, M. and Congleton, R.G., 1986, Accuracy Assessment: A User's Perspective. Photogrammetric Engineering and Remote Sensing, **52**: 397–399.

Thomas, N., Hendrix, C., and Congalton, R., 2003, A Comparison of Urban Mapping Methods Using High-resolution Digital Imagery. Photogrammetric Engineering and Remote Sensing, **69**(9): 963–972.

Tomlin, D., 1990, GIS and Cartographic Modeling, Prentice-Hall, Englewood Cliffs, NJ.

Townsend, P.A., 2000, A Quantitative Fuzzy Approach to Assess Mapped Vegetation Classifications for Ecological Applications. Remote Sensing of Environment, **72**: 253–267.

Veregin, H., 1995, Developing and Testing of an Error Propagation Model for GIS Overlay Operations. International Journal of Geographical Information Systems, **9**: 595–619.

Wang, F., 1993, A Knowledge-Based Vision System for Detecting Land Changes at Urban Fringes. IEEE Transactions on Geoscience and Remote Sensing, **31**: 136–145.

Wang, F. and Hall, G.B., 1996, Fuzzy Representation of Geographical Boundaries in GIS. International Journal of Geographical Information Systems, **10**(5): 573–590.

Woodcock, C.E. and Gopal, S., 2000, Fuzzy Set Theory and Thematic Maps: Accuracy Assessment and Area Estimation. International Journal of Geographical Information Systems, **14**(2): 153–172.

Worboys, M.F., 1998, Computation with Imprecise Geospatial Data, Computers. Environment and Urban Systems, **22**: 85–106.

Zhang, J. and Goodchild, M.F., 2002, Uncertainties in Geographical Information, Taylor & Francis, London.

Zhang, J. and Kirby, R.P., 2000, A Geostatistical Approach to Modeling Positional Errors in Vector Data. Transactions in GIS, **4**: 145–159.

Zhang, J. and Stuart, N., 2001, Fuzzy Methods for Categorical Mapping with Image-based Land Cover Data. International Journal of Geographical Information Science, **15**: 175–195.

INCOMPLETENESS, ERROR, APPROXIMATION, AND UNCERTAINTY: AN ONTOLOGICAL APPROACH TO DATA QUALITY

ANDREW U. FRANK
Institute for Geoinformation and Cartography, Technical University Vienna, Gusshausstrasse 27-29/127, A-1040 Vienna, Austria

Abstract. Ontology for geographic information is assumed to contribute to the design of GIS and to improve usability. Most contributions consider an ideal world where information is complete and without error. This article investigates the effects of incompleteness, error, approximation, and uncertainty in geographic information on the design of a GIS restricted to description of physical reality. The discussion is organized around ontological commitments, first listing the standard assumptions for a realist approach to the design of an information system and then investigating the effects of the limitations in observation methods and the necessary incompleteness of information. The major contribution of the article is to replace the not-testable definition of data quality as "corresponding to reality" by an operational definition of data quality with respect to a decision. I argue that error, uncertainty, and incompleteness are necessary and important aspects of how humans organized and use their knowledge; it is recommended to take them into account when designing and using GIS.

Keywords: incompleteness, error, approximation, uncertainty, error ontology, spatial ontology, spatial data quality

1. Background

The goal of human activities is to improve one's situation and – following the Golden Rule – to improve the "condition humaine" in general. This is part of a Greco-Judaic tradition to control the world and use it (Genesis 1, 28). Information became central for the development of economy in the past few centuries. The industrial revolution in the 18th and 19th centuries improved on the production of goods for human consumption and allowed

A. Morris and S. Kokhan (eds.), Geographic Uncertainty in Environmental Security, 107–131.

an unprecedented increase in population; it combined improvements in government, taxation, and markets together with technical improvements in manufacturing (North, 1981). North identifies a second economic revolution when scientific methods are used to produce systematically new knowledge to further advance technology and management. This is evident in the current debate on directing universities to produce "socially useful and responsible knowledge" combined with high levels of funding for universities but it is equally true for all the new Internet businesses. Information has become a factor of production, comparable to the classical production factors of land, capital, and labor (Ricardo, 1817, reprint 1996; Marx, 1867, translated reprint 1992; Frank, to appear 2005).

If information is a production factor like others, it must be measurable both in quantity and quality. Efforts to include "knowledge" in the accounting of large companies are under way (Schneider, 1996), but problems of measuring quantity and quality remain. Easily observable and countable substitutes (number of patents, number of scientific publications, etc.), which are expected to be proportional to the actual knowledge are often used. I have suggested a method to measure the quantity of pragmatic (useful) information (Frank, 2001; Frank, 2003b), but the approach is currently viable on a micro level only.

What do we mean when we say that information is of high quality? Before the computer age, one would have said "the information is from reliable sources," qualifying the information not directly but indirectly by its source. In today's information economy, quality of information becomes important for business. The loss for US businesses due to data quality problems is estimated as $600 billion for 2002 (Eckerson, 2006).

Quality of information is a novel concept, which has not been used before; scientists – especially astronomers and surveyors – commented on the quality of observations in the 18th century; surveyors have generalized this approach to evaluate the precision of observations and contributed to the data quality discussion (Chrisman, 1985; Frank, 1990; Robinson and Frank, 1985). Business processes using data go astray if the inputs are wrong and this gives an alternative approach to the topic (Wand and Wang, 1996).

Quality of information is even more important today in the transformation of the economy from maximum production at any cost to an economy respecting "limits of growth" (Meadows et al., 1972; Pestel, 1989). Mitigation of environmental disasters like flooding, tsunami, or forest fires are political and economical goals making detailed and high quality information necessary for their achievements. Mankind has learned that not everything that is technically possible is desirable. We need to understand the laws of environment as well as we understand the laws of physics: the construction of a perpetuum mobile (perpetual motion machine) is attempted today only by fools, because we understand the inviolable laws of thermodynamics. Plans affecting the environment too often violate environmental laws that are equally forceful; we find increasingly that "the environment kicks back" when we ignore its rules. Information, especially spatial information, plays the crucial role to understand and eventually use, to our advantage, the laws of environment.

2. Goal

In this article I want to link the methods used to collect, manage, and use environmental data with the ontological commitment, which are tacitly assumed. This seems useful to avoid that contradictions in the assumptions lead to confusion and inappropriate use of, in principle, valid methods. Identifying the ontological commitment is important in today's complex and diversified edifice of science to achieve consistency across different disciplines and applications. The focus is on data quality; the connection between data quality and ontology has been made before (Wand et al., 1996) and I hope to extend this original contribution in a way different from recent papers by (Ceusters and Smith, to appear 2006). The paper is restricted to descriptions of the physical reality and the extension of the arguments to cultural aspects, e.g., political boundaries, land ownership, etc. is left for future work.

The goal of Ontology in philosophy is to understand how the world is and how things exist in the world. It starts with Aristotle's Metaphysics (Aristotle, 1999). The term ontology was created in the 19th century; foundational contributions were made by Husserl. The difficulty with the philosophical tradition of Ontology is that human knowledge is limited to what we can observe; phenomenology concentrated on the limits of our abilities to know about the world (Bergson, 1896, reprint 1999). The movement of existentialism (Heidegger, 1927, reprint 1993; Sartre, 1943, translated reprint 1993) contrasts with analytical philosophy using increasingly formal methods, coinciding with foundational questions in mathematics (Whitehead and Russell, 1910–1913).

Ontology, as produced by philosophers, tries to give a consistent description of the world and how it exists in general. It is useful to identify the assumptions and point out inconsistencies but logical deduction alone cannot tell us "how the world is." It can at best, demonstrate that some set of assumptions are not consistent. Knowledge about the world is only achieved starting with observations and is thus dependent on the world *and* on the observation system.

Practical use of ontology – with a lower case o – in information systems has goals that are more modest: it gives rules how consistent descriptions of conceptualizations of a subset of reality for a purpose (Gruber, 2005). Any information system has an underlying ontology – the conceptualization of the part of the world, which is included – even if it is not described explicitly. Designers and users of an information system construct a mental model of the subset of the world they are interested in and communicate this model verbally; such symbolic representations are then entered in an information system. The data structure of the information system is a representation of this conceptualization; if it is described in a formal ontology (Smith, 1998) then the description can be analyzed and inconsistencies detected and resolved.

In this article, I want to explore the ontological commitments, which are necessary for an information system in a realistic view, i.e., a view that takes into account error, approximation, and uncertainty. The goal is to give a consistent set of ontological commitments, which allow a definition of data quality and how it is practically used. The focus is on geographic information systems and how they are used in environmental applications – but the results should be fully general for information systems independent of purpose. This approach is different from Wand and Wang's effort: they considered primarily questions of mapping between facts and their symbolic representation and assumed that the granularity of the representation is properly set; here I want to explore the difficulties that result from granularity and differences in the granularity when merging multiple data sets.

3. Ontological Commitments

Avoiding ontological commitments is not possible – designers of information systems can only avoid making explicit how they conceptualize the subset of the world they are modeling in the system. Making their choices explicit avoids inconsistencies, improves communication among multiple designers, and eventually communicates the conceptualization to the users of the system, again avoiding misunderstanding and misuse of the information provided by the system (Fonseca and Egenhofer, 1999). In this paper, I expand this view, which is common today, to include data quality aspects.

3.1. COMMITMENT O 1: A SINGLE WORLD

It is assumed that there is a physical world, and that there is only one physical world. This is a first necessary commitment to speak meaningfully about the world and to represent some aspects in a GIS. Few philosophers and many writers of science fiction (Asimov, 1957; Adams, 2002) have explored the logical consequences of constructions in which either no world outside of my thinking exists (Schopenhauer, 1819, 1844, translated reprint 1966) or in which multiple worlds coexist. They lead to inconsistencies, which often provide for interesting reading, but not to an account of the world as we experience it.

3.2. COMMITMENT O 2: THE WORLD HAS EVOLVING STATES

The world has states, which evolve in time. This ontological commitment is twofold: it posits a single time and changeable states of the world (this is postulate 1 of Wand and Wang (1996, 89)).

3.3. COMMITMENT O 3: OBSERVABLE AND CHANGEABLE STATES

The actors in the world can observe some of the states of the world at a given location and the present time (the *now* of Franck, 2004). Observation of physical state for certain properties and a point is objectively possible (point observations); the influence of the observer on the observation value is small and repeated observations give the same values.

An extensive discussion of the influence of the observer on the observation has been carried out in the social sciences, where subjective judgments of situations are heavily influenced by the background of the observer and in physics, where observation influences the state of the observable (Leinfellner, 1978; Mittelstraß, 2003). These difficulties do not affect the treatment here; the observable states of the world are restricted to states of the physical (macro) reality as they are measured with standard measurement devices and the result expressed in SI units or similar (e.g., cm, g, s). I exclude from this discussion (a) physiological states of individuals, as discussed in measurement sciences (Krantz et al., 1971); (b) assessment of cultural reality are included in the observable states; nor (c) quantum physics. The observable states of importance in a GIS are within the realm of classical physics and do not include quantum effects.

The actors in the world can not only observe the world, but also they can affect changes in reality through actions. The effects of actions are changed states of the world and these changed states can be observed. This gives the *semantic loop* (Figure) that connects the observations with their sensors to the changes with their actuators and combines the semantics of observations with the semantics of changing operations (Frank, 2003a).

3.4. COMMITMENT O 4: INFORMATION SYSTEMS ARE MODELS OF REALITY

Observation results in information and we have to discuss both the system of reality and the information system (postulates 2 and 3 in Wang and Wand (1996, 90)). Observations translate the state of the world from the realm of reality to the realm of information (Figure 1). The information realm is a partial and incomplete model of the world, somewhat as described by Plato in his cave metaphor. By model, we understand a structure, which is related by a morphism with the world. Corresponding operations in the model have corresponding results (Kuhn and Frank, 1991; Goguen and Harrell, 2006). The focus of Wang and Wand is this mapping, which they characterize along the same lines as customary in category theory (Asperti and Longo, 1991), (similar recently Ceusters and Smith, 2006).

The division of the ontology in a world realm and an information realm is an important contribution of Wand and Wang; the two realms are related by a morphism. The quality of the information is threatened if the relation between the things in the world and the entities in the information realm are not isomorphic (i.e., one–one). If the two realms are linked by an isomorphism, which is often assumed, then the distinction between thing in the world and in the information model would not be required from an algebraic point of view. Reality and the model are the same – up to isomorphism (Mac Lane and Birkhoff, 1991). Actual information systems do not achieve an isomorphism: sometimes one thing in the world

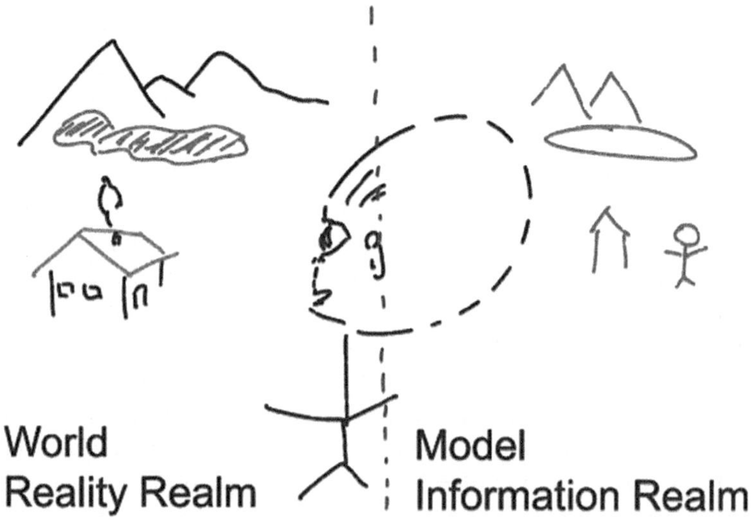

Figure 1. The reality realm and the information realm

corresponds to multiple entities in the information model, sometimes situations in the world have no related representation in the information model; in most cases, the simplification resulting from the assumption of an isomorphism is not justified and the mapping between reality and information model must be analyzed and not be glossed over (Kent, 1979).

3.5. COMMITMENT O 5: SEPARATE PHYSICAL AND INFORMATION CAUSATION

The changes in the state of the world are modeled by physical laws: The cause for water flowing downward is gravity, the cause of a bullet to fly are the forces resulting from a chemical reaction, when the explosive in the shell is ignited. The rules of physics can be modeled in the information realm and allows the construction of future states in the information realm. This is extensively used to predict what the effects of actions are and the foundation of all planning. The change in the physical world can be modeled as a Markov chain – a following state is the result of the current state or of the current state and previous states.

A second and entirely different form of causation, which I will call *information causation,* is the result of decisions by agents. Agents have *free will* and can make decisions about their actions (Searle, 2001). Decisions are in the information realm but they affect – through physical laws – the reality realm. In a macroscopic view, a successor state is independent of the previous state of the world.

It is however, important to note that decisions can have the intended effect if – and only if – the action can be carried out and no physical laws contradict it. For example, deciding to move from Vienna to Kiev in one hour by car is possible, but the decision cannot be carried out because several physical (and traffic) laws prohibit my car to drive at the necessary speed of 1,000 km/h, etc. Despite my decision, the desired result cannot be achieved, because I cannot start a chain of physical causations to realize my decision; one could say that the mapping of the information causation from the information realm to the reality realm does not exist in this case.

4. Quality of Information

How to define quality of information in this context? How to give more content to the idea that given information is of high quality if it corresponds with reality? Indeed, what is meant by "correspond?"

The most often used definition of information quality is based on the repeatability of observations. Assuming that a state of the world has not changed, then an information is correct in this sense that if the information is the same as obtained by a new observation. This is the definition used by Wang and Wand. In their contribution, they point out that not only the recorded information must be considered, but also other information that can be inferred (definition 2 and postulate 6). This definition assumes that the quality of the information is – at least in some dimensions – fixed appropriately with respect to the intended use. In a world that is constantly changing, observations cannot be exactly repeated – an observation made later is different from the observation made before; the customary definition is usable only if these effects are ignored and thus, strictly speaking, only a definition for "correctness with some limits."

This more traditional approaches to data quality is often used, because it considers the production of the data and deduces the quality of data as properties resulting from the production process (Timpf et al., 1996; Timpf and Frnak, 1997; Timpf, 2002). This is similar to the view that the quality of a car results from the details of the production process and thus fits general methods to assess product quality. Unfortunately, these definitions of data quality are mostly irrelevant for geographic data and its use. Practitioners resist to use data quality descriptions following current standards (Hunter and Masters, 2000) because they are not informative for potential users.

An alternative definition is based on the concept of "fitness for use" (Chrisman, 1985). The information is used to make decisions, which are then translated into actions. This is the only use of information and is reflected in a convenient definition of information: Information is an answer to a human question (Frank, 1997). People ask questions in order to make decisions, sometimes the decisions are imminent and sometimes we just collect information to be prepared for later decisions. If information is used to make decisions, then the quality of the information can be related to the quality of the decisions made.

To assess the quality of the decision brings us back to the semantic loop: reality and information realm are connected by (1) observations, which populate the information realm and (2) the decisions and actions, which change the world (Figure 2). To assess the quality of the information one must assess the quality of the decision and how it is influenced by the information. The connection between data semantics and quality is revealed with this viewpoint.

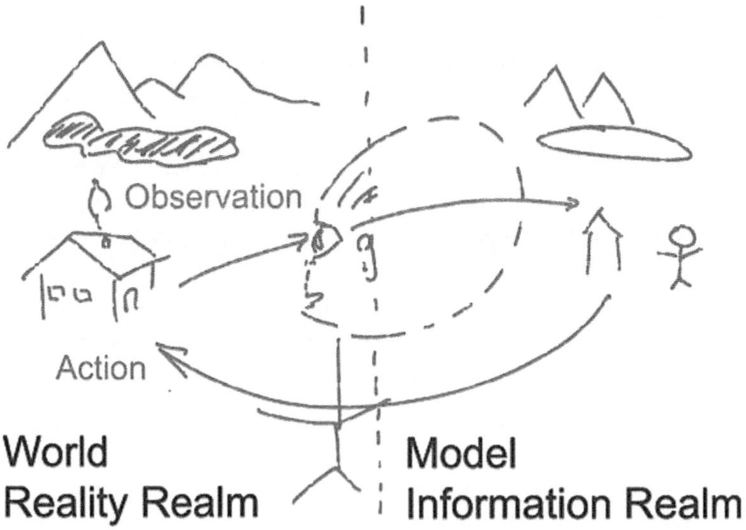

Figure 2. Closed loop semantics connect reality realm with information realm through observations and actions

5. Ontology of Error and Uncertainty

The ontological commitments were listed above primarily to remind the readers of what is implied in an information system design; they are the usual assumptions underlying the construction of geographic information systems (Frank to appear) and are the justification for the ontology-driven approach to design an information system (Fonseca and Egenhofer, 1999). These "usual" ontological commitments are unrealistic, because they ignore error and uncertainty in our knowledge of the world, and pretend that we have perfect knowledge. This illusion is acceptable given that we have most of the time sufficient information to function at acceptable performance levels in our environment but it is not usable to construct more advanced information systems, which use data collected for other purposes and following different quality standards; in such combinations of data from different sources, we must take into account the limitations in our knowledge. Understanding error and uncertainty in data is therefore crucial to achieve interoperability of geographic data collections (Vckovski, 1997).

5.1. COMMITMENT EU 1: INFORMATION ABOUT THE WORLD IS INCOMPLETE

The world is infinitely complex and the information we have about it is always limited. It is impossible to construct a fully accurate and detailed model of the world, because such a model would be at least as big as the world!

The information model of the world we construct is therefore always limited in the level of detail and the completeness of the aspects modeled; most of what is in the world must be left out; our models are restricted to the aspects that are relevant for the decisions we intend to make. The level of detail is linked to the purpose of the GIS and the decision expected to be made with the information. This contradicts the optimistic view of GIS as a single "multi-purpose" or even "all-purpose" spatial information system in the early days (Gurda et al., 1987).

5.2. COMMITMENT EU 2: OBSERVATIONS ARE ERRONEOUS

Observations of the changeable states of the world are never perfect. They are affected by unavoidable effects, which create differences between the ideal observation (the ideal *true value*) and the actual realization of the observation. These effects can be random and are often modeled by normal distributions. Observations are also affected by systematic effects, e.g., a yardstick is too short or a watch runs always slow. Such systematic effects can be controlled and eliminated by observation methods, but random disturbances cannot be avoided and affect all observations of physical properties.

5.3. COMMITMENT EU 3: AUTOCORRELATION

One might ask how people survive in a world where the information we have is necessarily incomplete and erroneous. To conclude that goal-directed actions and survival is impossible, would be premature ("Philosophers should be very careful when they deny the obvious," Searle). But what counteracts the effects of the fact that all our knowledge is incomplete and erroneous?

The physical world is strongly autocorrelated – both in space and time. The most likely observation of a property just a little bit to the left where I looked before, or just a little bit later is most likely very similar to the observation I made before. Correlation between different observations is also strongly correlated: for example sugar content of a fruit and its color is

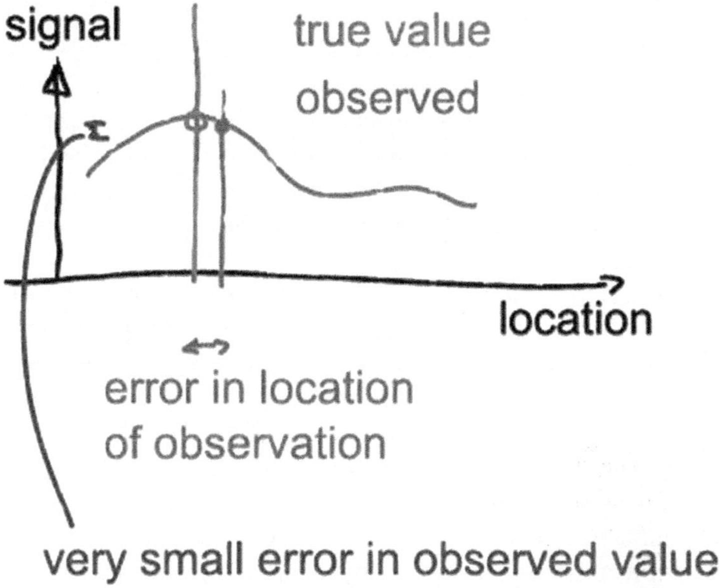

Figure 3. Large error in location leads to small error in observed value

often correlated – we pick the red strawberries, because they are sweet and taste good and leave the green ones. The strong autocorrelation is also at the base that the usual definition of data quality as "corresponding to reality" works. An observation a little bit later or nearly at the same place produces nearly the same value; repeated observations would not be meaningful in a world not strongly, spatially, and temporally autocorrelated (Figure 3).

Life in a world without the strong spatial and temporal autocorrelation would be very difficult if not impossible. Most of the world is slowly and continuously changing and we focus on the discontinuities. On the background of stability we focus on the interesting, changing, and discontinuous points.

5.4. COMMITMENT EU 4: BIOLOGICAL AGENTS HAVE LIMITED INFORMATION-PROCESSING ABILITIES

The structure of our information is not only influenced by reality but also by the systems to process information. The abilities of the brains of biological agents – including humans – are very limited and the biological, i.e., energy, cost of information processing, is high. Biological agents have therefore developed methods to reduce the load on their information processing systems – commonly called the "cognitive load" – to allow efficient decision-making with limited effort and often in short time.

5.5. COMMITMENT EU 5: OBJECT-CENTERED DATA PROCESSING

Processing of information describing reality is primarily object oriented. Humans, and many other biological agents, structure the observations they perceive in information about objects. The observation of properties of points in space and time are restructured to become properties of objects. Objects are constructed such that they endure in time and have constant properties over time. Spatial and temporal autocorrelation makes this reduction of cognitive load possible.

The cognitive system forms objects at boundaries of continuities and reduces therewith the cognitive load; it is simpler to keep track of objects with uniform and seldom changing properties and to pay attention to their boundaries; most of the world modeled as objects is stable, uniform, and unchanging compared to a point (raster) model of the world. Autocorrelation is similarly used in technical systems to reduce bandwidth necessary for transmission, e.g., of television images; it is the reason data compression methods like JPEG and MPEG work.

Our cognitive system is so effective because it identifies objects in the array of sensed values, and we reason with objects and their properties, not with the multitude of values sensed. Thinking of tables and books and people is much more effective than seeing the world as consisting of data values for sets of cells. It is economical to store properties of objects and not deal with individual raster cells. We cut the world in objects that are meaningful for our interactions with the world. As John McCarthy has pointed out:

> [S]uppose a pair of Martians observe the situation in a room. One Martian analyzes it as a collection of interacting people as we do, but the second Martian groups all the heads together into one subautomaton and all the bodies into another. ... How is the first Martian to convince the second that his representation is to be preferred? He would argue that the interaction between the head and the body of the same person is closer than the interaction between the different heads ...when the meeting is over, the heads will stop interacting with each other but will continue to interact with their respective bodies." (McCarthy and Hayes, 1969, 33)

Our experience in interacting with the world has taught us appropriate subdivisions of continuous reality into individual objects. Instead of reasoning with arrays of connected cells, as it is done in finite element analysis, e.g., strain analysis or movements of oil spill, reasoning is performed with individual objects: The elements on the tabletop (Figure 4) are divided in objects at the boundaries where cohesion between cells is low; a spoon consists of all the material that moves with the object when I pick it up and move it to a different location.

Figure 4. Typical objects from tabletop space

Humans conceptualize themselves and the rest of the reality preferably in terms of objects and their properties. Objects endure in time, they have an identity and changeable state. The changeable state of objects is the consequence of the assumption that the world has changeable states and objects are aggregates of real world points. Object properties describe the state of objects; they are typically integrals over the volume the object forms (Equation 1). This is usually tacitly assumed (e.g., in Wand and Wang) but creates ontological difficulties:

object formation is not unique: different persons and for different purposes the same part of reality can be split in different ways into objects.

The formation of object introduces error and uncertainty in the data

$$P(O) = \iiint_{V(O)} p(v)\, dV \qquad (1)$$

5.6. COMMITMENT EU 6: MULTIPLE WAYS TO FORM OBJECTS

Aristotle discussed familiar objects in terms of natural kinds – the classes of objects that are naturally distinct: cats, dogs, etc. There is little doubt how to form such objects and how to classify them for the natural species, because

Figure 5. Three girls combing a big dog, making the boundaries of the dog sharper

there exist hardly any borderline cases – there are no breeds between dogs and cats (but not all cases are as simple: horses and donkeys breed and produce mules). The task of the philosopher is to cut up nature at its joints. An object is considered to move as a single unit: a glass, a plate, a cat. All that moves with the object is part of the object – and only exceptionally one asks question like "are loose hair in the fur of an animal part of the animal or not?" (Figure 5).

Object formation is however not as simple when we consider geographic space: there are multiple ways to subdivide space into objects. Considering the terrain, we can focus on form, and identify watersheds, valleys, and mountains, but focusing on land cover, we identify fields and forests. Many other ways to subdivide space are used: ethical and religious boundaries are often debated and sometimes lead to wars. For a geographic information system, we must accept that not a single "natural" subdivision of space exists but different purposes require different approaches; a GIS must be prepared to have coexistent, overlapping spatial objects.

5.7. COMMITMENT EU 7: OBJECTS AS REGIONS WITH UNIFORM PROPERTIES

A very general approach to define objects is to say that they form regions with at least one property having a uniform value. The prototypical object – an animal – is then a connected area of cells with the same DNA; other objects are uniform in material, color, movement, etc. Solid objects, where boundaries are revealed when we move them, have uniformity in material properties that makes them "hang together."

For properties with a continuous value, the uniformity of the property means to be within some limits and introduces thresholds for which the property is considered uniform. This absorbs uncertainty in the observations but introduces uncertainty in the boundary of the object.

Different objects result if we select different uniform properties. Areas of uniform land cover (e.g., grass land, forest) do not necessarily coincide with watersheds and produce different objects. The autocorrelation in space and the correlation between factors influencing natural processes result in object boundaries that often (nearly) coincide. It is not by accident that the land cover on one side of the fence is different than on the other side and that the boundary of "my garden" and the neighbor's field coincide with the fence.

5.8. COMMITMENT EU 8: OBJECT FORMATION IS UNCERTAIN

Objects are delimited by boundaries and these boundaries have observational error; a general model of objects defines them as areas of uniform values in some property. The error in observing the property value affects the determination of the boundary (Figure 6).

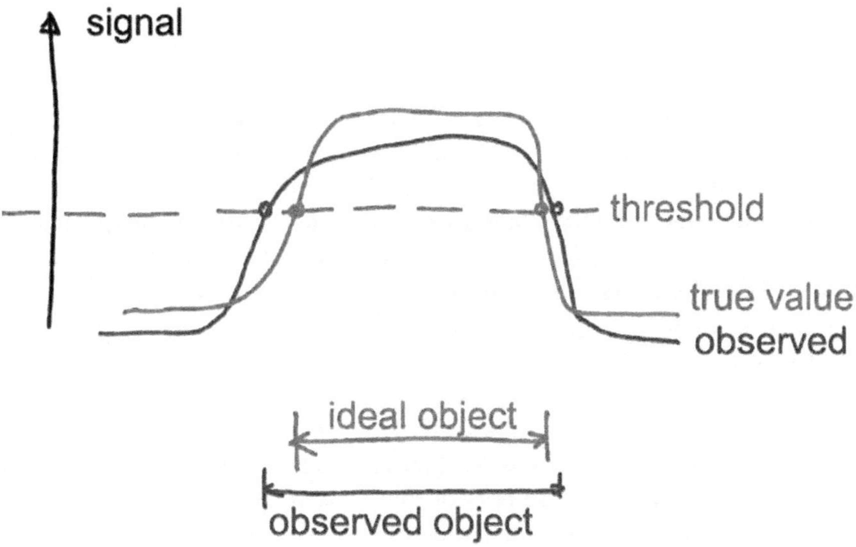

Figure 6. Error in observation of property results in error of object boundary

The objects have states that derive from the observations of point properties (commitment O 3). The properties of objects are sums (integrals) over some functions of point observations. The error in the observation of properties of the objects is therefore affected by observational errors in multiple forms:

Error in the area,
Error in the observation

These errors can be modeled if the observation errors are assumed to be random, normally distributed (Navratil and Frank, 2006). However, such simplifying assumptions that are necessary to achieve a tractable formalization are unrealistic as they leave out the influence of correlations. I suggest using the term *approximation* for the difference between the true, intended value and the value resulting from observations of properties of objects.

5.9. COMMITMENT EU 9: UNCERTAINTY IN CLASSIFICATION

Objects are not just formed and described, but the formation and the description is detail to the classification of a phenomena as an object of a certain type. Objects are first instances of a class – even if this is only the most general class *Thing* – and then boundaries and properties are observed. This classification of an object has some problems that affect the quality of the data as we will see after a brief discussion of the concept of *class* (also known as *universal*, *type*, etc.) and how it is used in decision-making.

The classification asserts that the object in case is part of a group of objects – the class – that share some properties. There are two ways classes are defined. In the extensional understanding of a class it is a set of objects with common properties; the intensional definition of class starts with the properties of an object (or its intended use) and the class is all objects (existing, having existed, or existing in the future) with these properties. One can imagine an ideal member of this class – the prototypical dog, mountain, etc., which is imagined as ideal universal, akin to the ideal circle; philosophers debate how such universals exist. The practical problem for information systems is that different definitions of classes are used, but described by the same word.

The descriptive terms for classes (forest, lake, etc.) are often polysemous – there are multiple concepts described by the same term. In Austria, the word "forest" is used with different meanings, some of them apply even when no trees are present (but also, some terrains with trees do not classify for "forest" in a legal sense).

Classification is further complicated by the "prototype effects" in natural language classification: not all objects in a class have some properties in common (Rosch, 1978). Take the example of the class "bird"; one would commonly assume that birds are animals that fly – just to be reminded that also ostriches, emus, and penguins are birds, which cannot fly. Some exemplars are just better "birds" than others. This applies equally to land use and land cover classifications; the prototypical forest in central Europe (the "dunkle Tann" occurring in Grimm's fairy tales) is different from what a Greek or a Finn calls a forest.

Classifications evolve in time with advances in science (Fleck, 1935, reprint 1980) or with changes in the social interest. For example, land cover definitions evolved in time and the observations made based on a previous classification are incommensurable to observations with the new classification (Comber et al., 2004).

Classification is very important for human communication: we speak of cats and dogs and mean the classes of animals that have particular properties, e.g., size, form, behavior. Classifying an object based on its visible properties leads us often to assume that the object has the values typically for objects of this class for properties that we cannot observe; for example, if we classify an animal as a dog based on its visual appearance we will assume that it barks (and be very surprised if it starts to meow). Classification is thus the base of "default reasoning" when we do not have particular information about the individual we assume that the usual properties of the class apply.

The uncertainty in classification comes from multiple sources, including at least:

Selection of the property, which is uniform in the object
Selection of the thresholds for uniformity
Error in the position of the boundary
Errors in the observations relevant for the subclassification of an object

Here an example: for land use classification, the property that must be uniform for an object is the land use (not land cover – but given that land cover is easier to observe, most classifications of land use are actually classification of land cover). Depending on the scale of our mapping efforts, wider or narrower thresholds for "uniformity" are set: how much weeds may grow in a corn field before we stop classifying it uniformly as "corn." How fine are the subdivisions for land use: agricultural (vs. forest), field (vs. pasture), corn field (vs. wheat field). Once we have settled on corn fields and set the thresholds for weeds, the boundary of the field must be determined and measured. If we then further classify in corn fields of high yield and corn fields of low yield, an estimate of the yield is necessary.

Under the assumption that a classification process groups objects based on some determined properties in groups, the uncertainty in the classification would be only from the error in the observation of properties. The formation of objects involves the uncertainty of the boundary and the errors in property observation. The approximation in the object property translates to an uncertainty in the classification. If a more reliable and precise classification is available, then the quality of a given classification can be assessed and the percentage of omissions and commissions established or a matrix of misclassification of multiple classes given. The difficulty is however more often in the imprecise or changed definition of the classes, which makes object formation and classification nearly impossible to compare (Comber et al., 2004).

The uncertainties in classification are multiple and poorly understood. Many ontologists posit that classes with fixed definitions exist, ignoring that many of the usability problems of information systems originate in differences in the classifications used during data collection and data use. I have suggested that properties of objects are used as primitive notions (and not classes as usual in taxonomies) and that classifications are defined in terms of object properties; this results in very fine-grained classifications and defined rules of inference between classes (Frank, to appear 2006a, b; the idea is related to Formal Concept Analysis (Burmeister, 2003).

6. Decision Process

The commitments to incomplete, uncertain, and erroneous information must now be linked to the decision process to see how they affect the quality of the decisions. This requires a summary model of how decisions are taken:

The decision to take some actions starts with a goal, an imagined future world state that is desirable to the agent. For example, I am hungry and imagine a future world state in which I have eaten. I consider then a set of alternative actions to achieve that state and evaluate the different plans in order to select the best course of action, which I then carry out. Not all aspects of this model must be conscious to the agent – it is sufficient that the agent selects one of the alternatives because it appears – given the current state of his knowledge – the best option. It is implied that decisions can be wrong and that decisions are made with insufficient information, etc. The decision is sufficing and the rationality is bounded by the limitations of the agent (Simon, 1956).

7. Correct Decisions Derive from the Quality of the Information

Information cannot be correct in the sense of correspondence with reality (commitment EU 1 and 2): a repeated observation is never giving exactly the same result; the random effects and the changes in reality produce different values every time the observation is repeated. Statistical tests can be used to assess if the new value obtained is within the expected margins with a certain probability.

A practical definition is to state that information is correct if it leads to correct decisions. This requires first a definition of what we mean by "correct decision." Let me start with a counterexample: information is incorrect if it leads to a wrong decision. For example, my decision to go to the airport at 7:30 a.m. to catch the plane for Frankfurt is in error if the plane has actually left at 7:15 a.m. Other example: my decision to buy 2 m extension cord to connect my stereo system is incorrect if I find at home that the cable is too short because the distance between the power outlet and the plug is 3 m. A decision is not correct if it does not lead to the desired goal (i.e., flying to Frankfurt, connecting the stereo set) – this points out that decisions are taken in order to achieve a certain goal; if the action decided upon does not lead to the desired goal, the decision is incorrect.

If we assume (bounded) rationality in the decision process, the information available is influencing the decision – thus information that leads to the correct decision is correct information. Note that this definition does require much less, than the definition of correctness based on repeatability and takes into account the influence of error and uncertainty on the information. Much error, uncertainty, and incompleteness in the information can be tolerated as long as the action decided on achieves its goal. A decision can be wrong in multiple ways:

The action that is decided cannot be carried out.
The achieved state of the world does not satisfy the goal.
The action was not optimal; if the information would be better, another action would have been selected.

It appears useful to analyze these different reasons for actions to fail the achieved goal.

7.1. PHYSICAL IMPOSSIBILITY DUE TO OBSERVATION (MEASUREMENT) ERROR

An action is not possible because of observation errors. This is the type of error extensively studied by surveyors: Most spectacular are the measurements taken to assure that the two ends of a tunnel meet in the middle of the mountain. Similar cases of careful measurement, a surveyor measures the gap between the roads on both sides of the river and measures the steel bridge, which should fit in the gap. If the bridge is too long or too short, closing the gap is not possible.

In general, humans have found methods to avoid such costly and difficult measurements that have always some error. Carpenters traditionally put the beams together, cut and bore the holes at once through multiple layers and thus assure that the pieces will fit when installed in the roof – all without measurement! If a cable of a certain length is necessary most people do not try to cut to measure but make it longer – it will fit certainly, even if measurement errors are relatively large (I should have bought a 5 m extension cord – it would have achieved my goal with a small additional cost!).

Many such techniques have been devised over the millennia of carpentry, tailoring, etc. to reduce the need for exact measurement; most trades avoid measurements completely! Only few situations make surveying and exact measurement necessary, for example, the reconstruction of a boundary after it is lost due to flooding in Ancient Egypt. Measurements are necessary, when there is no "sure side" where error does not matter: A cable can be too long without problem, a box can be too large to pack an object, but some problems have no "secure side" – too long or too short is equally bad. For example, cooking pasta or baking bread requires exact timing – but again the goal is achieved by repeated testing and not by accurate measurement. Advanced technology increases the need for accurate measurement and planning – sea navigation, building construction with accurate planning of the forces in the building and reduced, slender pillars and many similar modern examples are only possible with accurate measurement and precision of measurement are taken into account in the design.

7.2. PHYSICAL IMPOSSIBILITY DUE TO LACK OF KNOWLEDGE

An action can be impossible because some crucial information was not available. For example, driving to a city and finding out that the city is on the other side of a river or on an island – in both cases a means to cross the river (a bridge, a ferry) is required. A case of an instruction from a car navigation system to cross a river, where a ferry should be used and was not present was widely publicized, because the driver drove the car into the river and blamed the incomplete information from his navigation system (Raubal and Kuhn, 2004). This case of omission is of great importance and it is much more difficult to guard against it; Grice with his conversational implication studied information and decisions in the context of a exchange between people, but the theory is applicable to the information we gain from consulting a database (Grice, 1989).

7.3. THE ACTION SELECTED IS NOT OPTIMAL

Information present is incorrect and therefore the selected action is not optimal for the situation; this is often a case of a commission error: a map shows a road, which is not (yet) existing and one decides on a short route, which later is discovered to be longer than another route.

The economic effects are in general not very important – because the difference between optimal choice and second, third best choice are not large. This is an effect of the autocorrelation already mentioned but also part of the intentional construction of infrastructure in the world that are whenever possible redundant – if one fails, there is always a second option. Mankind has learned how to live in a world of error and uncertainty!

7.4. ERRORS IN ROAD NAVIGATION DECISION

In a decision on road navigation, that is, which road to follow to drive to another place, the three types of errors in decisions due to information quality can be explained. Assume that we need to drive on a Sunday from *A* to *B* and have gas in the car for 100 km; the information we have is shown in Figure 7 (left). The shortest path seems to be *x*. This decision is in error due to imprecise measurement, if the path *x* is very convoluted and actually 120 km long and we will fail to reach our goal. The decision to follow path *z* is in error for lack of knowledge that *B* is on an island and the ferry runs only at workdays (on Sundays one should take path *y*). The decision to take path *x* is not optimal if we find out that the length of path *z* is not 85 km as marked but only 65 km; it would have been a better decision to take *z* and not *x*.

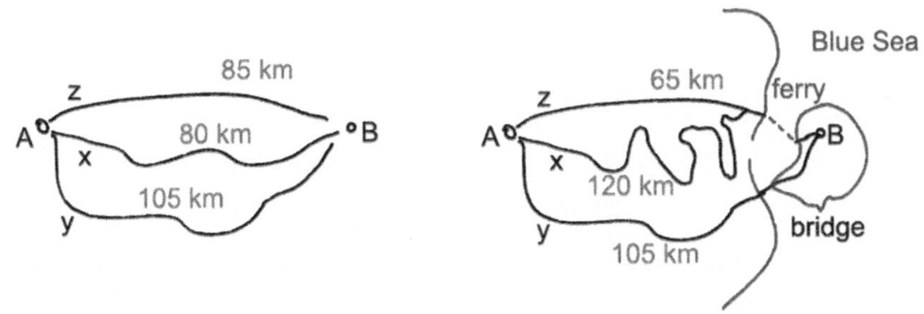

Figure 7. Information available for decision and true situation

8. Conclusion

The economic effects of measurement errors and commissions are often not very important, but errors of omission are difficult to counteract and have substantial cost. This may give a partial reason why people collect information "just in case." Who has not a large library and reads all papers published in the hope that the data obtained may be useful one day? In general, the information we have is sufficient for the decisions we must make and information errors are not very costly, but often we lack the information necessary completely.

Geographic data used for administrative decision-making is usually collected with proper levels of quality to make the intended decisions "reasonably well." By reasonably well, I mean that an optimum is reached between the cost of improved data quality through more efforts when collecting the data and the cost of correcting errors in the decisions due to errors in the data (disregarding situations where low data quality is favoring one politically influential group over another and low data quality is therefore politically desirable).

If geographic data is used for purposes it was not originally intended, for example, using administrative data for environmental planning, the particulars of the quality of the data for this decision must be considered carefully.

In this contribution I have tried to show the effects of observation errors and how they lead to approximation of value describing objects and result in uncertainty in the classification. This does not give a set of dimensions for data quality, as has been attempted before (Chrisman, 1985; Frank, 1990; Wand and Wang, 1996); efforts to identify dimensions of data quality seem not to avoid the correlation between different dimensions: temporal or spatial resolution cannot be separated in two independently observable dimensions (and similarly for other dimensions). At the present time, I note simply that a definition of separable dimensions of data quality cannot be achieved.

References

Adams, D. (2002). *The Ultimate Hitchhiker's Guide to the Galaxy*. Del Rey.

Aristotle (1999). *Metaphysics*. Penguin Classics.

Asimov, I. (1957). *Earth is Room Enough*. NY, Doubleday.

Asperti, A., and G. Longo (1991). *Categories, Types and Structures – An Introduction to Category Theory for the Working Computer Scientist*. Cambridge, MA, The MIT Press.

Bergson, H. (1896; reprint 1999). *Matière et Mémoire. Essai sur la relation du corps et l'esprit*. Paris, Les Presses Universitaires de France.

Burmeister, P. (2003). Formal Concept Analysis with ConImp: Introduction to the Basic Features. Darmstadt, Germany, TU-Darmstadt: 50.

Ceusters, W., and B. Smith (2006). *A Realism-based Approach to the Evolution of Biomedical Ontologies*. Forthcoming in Proceedings of AMIA 2006, Washington, DC.

Ceusters, W., and B. Smith (to appear 2006). *Towards A Realism-Based Metric for Quality Assurance in Ontology Matching*. FOIS, Baltimore, MD.

Chrisman, N. (1985). An Interim Proposed Standard for Digital Cartographic Data Quality: Supporting Documentation. In: H. Moellering (ed.). *Digital Cartographic Data Standards: An Interim Proposed Standard.* Columbus, OH, National Committee for Digtial Cartographic Data Standards, 6.

Comber, A., P. Fisher, and R. Wadsworth (2004). In: A. D. Bruck (ed.). *Comparing of Expert Relations Between Land Cover Datasets*. ISSDQ'04, Leitha, Austria, Department of Geoinformation and Cartography.

Eckerson, W. W. (2006). Data Warehousing Special Report: Data Quality and the Bottom line. Retrieved August 08, 2006 from http://www.adtmag.com/article.aspx?id=6321&page.

Fleck, L. (1935; reprint 1980). *Entstehung und Entwicklung einer wissenschaftlichen Tatsache. Einführung in die Lehre vom Denkstil und Denkkollektiv*. Frankfurt a. Main, Suhrkamp.

Fonseca, F. T., and M. J. Egenhofer (1999). *Ontology-driven Geographic Information Systems*. 7th ACM Symposium on Advances in Geographic Information Systems, Kansas City, MO.

Franck, G. (2004). Mental Presence and the Temporal Present. In: G. G. Globus, K. H. Pribram and G. Vitiello (eds.). *Brain and Being: At the Boundary Between Science, Philosophy, Language and Arts*. Amsterdam, Philadelphia, John Benjamins: 47–68.

Frank, A. (to appear 2005). A Case for Simple Laws. In: B. Smith, I. Ehrlich and D. Mark (eds.). *The Mystery of Capital and the New Philosophy of Scocial Reality*, 288.

Frank, A. (to appear 2006a). Distinctions – A Common Base for a Taxonomic Calculus for Objects and Actions. *Spatial Cognition and Computation.*

Frank, A. (to appear 2006b). *Distinctions Produce a Taxonomic Lattice: Are These the Units of Mentalese?* International Conference on Formal Ontology in Information Systems, Baltimore, MD.

Frank, A. U. (1990). Qualitative Spatial Reasoning about Cardinal Directions, University of Maine, NCGIA.

Frank, A. U. (1997). Spatial Ontology: A Geographical Information Point of View. In: O. Stock (ed.). *Spatial and Temporal Reasoning*. Dordrecht, The Netherlands, Kluwer: 135–153.

Frank, A. U. (2001). The Rationality of Epistemology and the Rationality of Ontology. In: B. Smith and B. Brogaard (eds.). Rationality and Irrationality, Proceedings of the 23rd International Ludwig Wittgenstein Symposium, Kirchberg am Wechsel, August 2000. Vienna, Austria, Hölder-Pichler-Tempsky, 29.

Frank, A. U. (2003a). Ontology for Spatio-Temporal Databases. In: M. Koubarakis, T. Sellis (eds.). *Spatiotemporal Databases: The Chorochronos Approach*. Berlin, Springer: 9–78.

Frank, A. U. (2003b). Pragmatic Information Content: How to Measure the Information in a Route Description. In: M. Goodchild, M. Duckham and M. Worboys (eds.). *Perspectives on Geographic Information Science*. London, Taylor & Francis: 47–68.

Frank, A. U. (to appear). Ontology for GIS. Vienna, Technical University Vienna, Institute for Geoinformation and Cartography.

Goguen, J., and D. F. Harrell. (2006). Information Visualization and Semiotic Morphisms. Retrieved September 01, 2006, from http://www.cs.ucsd.edu/users/goguen/papers/sm/vzln.html.

Grice, P. (1989). *Studies in the Way of Words*. Cambridge, MA, Harvard University Press.

Gruber, T. (2005). TagOntology – a way to agree on the semantics of tagging data. Retrieved October 29, 2005, from http://tomgruber.org/writing/tagontology-tagcapm-talk.pdf.

Gurda, R. F., D. D. Moyer, B. J. Niemann, and S. J. Ventura (1987). *Costs and Benefits of GIS: Problems of Comparison*. International Geographic Information Systems (IGIS) Symposium (IGIS'87), Arlington, Virginia.

Heidegger, M. (1927; reprint 1993). *Sein und Zeit*. Tübingen, Niemeyer.

Hunter, G. J., and E. Masters (2000). *What's Wrong with Data Quality Information? Abstracts*. International Conference on Geographic Information Science, Savannah, GA.

Kent, W. (1979). *Data and Reality Basic Assumptions in Data Processing Reconsidered*. Amsterdam, New York, Oxford, North-Holland.

Krantz, D. H., R. D. Luce, P. Suppes, and A. Tversky (1971). *Foundations of Measurement*. NY, Academic Press.

Kuhn, W., and A. U. Frank (1991). A Formalization of Metaphors and Image-Schemas in User Interfaces. In: D. M. Mark and A. U. Frank (eds.). *Cognitive and Linguistic Aspects of Geographic Space*. Dordrecht, The Netherlands, Kluwer, 419–434.

Leinfellner, E. (1978). *Ontologie, Systemtheorie und Semantik*. Duncker & Humblot GmbH.

Mac Lane, S. and G. Birkhoff (1991). *Algebra Third Edition*. Providence, RI, AMS Chelsea.

Marx, K. (1867; translated reprint 1992). *Capital: Volume 1: A Critique of Political Economy*. Pengiun Classics.

McCarthy, J., and P. J. Hayes (1969). Some Philosophical Problems from the Standpoint of Artificial Intelligence. In: B. Meltzer and D. Michie (eds.). *Machine Intelligence 4*.Edinburgh, Edinburgh University Press: 463–502.

Meadows, D. H., D. l. Meadows, J. Randers, and W. W. Behrens III (1972). *Limits to Growth*. NY, Universe Books.

Mittelstraß, J. (2003). *Transdisziplinarität – wissenschaftliche Zukunft und insitutionelle Wirklichkeit*. Uvk.

Navratil, G., and A. Frank (2006). *What Does Data Quality Mean? An Ontological Framework*. AGIT 2006, Salzburg, Austria, Wichmann Verlag.

North, D. C. (1981). *Structure and Change in Economic History*. New York, London, W. W. Norton.

Pestel, E. (1989). *Beyond the Limits to Growth: A Report to the Club of Rome*. NY, Universe Books.

Raubal, M., and W. Kuhn (2004). Ontology-based task simulation. *Spatial Cognition and Computation* **4**(1): 15–37.

Ricardo, D. (1817; reprint 1996). *Principles of Political Economy and Taxation*. Prometheus Books.

Robinson, V. B., and A. U. Frank (1985). *About Different Kinds of Uncertainty in Collections of Spatial Data*. Seventh International Symposium on Computer-Assisted Cartography, Auto-Carto 7, Washington, DC, ASP and ACSM.

Rosch, E. (1978). Principles of categorization. In: E. Rosch and B. B. Lloyd (eds.). *Cognition and Categorization*. Hillsdale, NJ, Erlbaum.

Sartre, J. P. (1943; translated reprint 1993). *Being and Nothingness*. New York, Washington, Square Press.

Schneider, U., ed. (1996). *Wissensmanagement – Die Aktivierung des intellektuellen Kapitals*. Frankfurter Allgemeine Zeitung GmbH.

Schopenhauer, A. (1819 & 1844; translated reprint 1966). *The World As Will and Representation (Volume 1 & 2)*. Dover Publications.

Searle, J. R., ed. (2001). *Rationality in Action*. MIT Press.

Simon, H. (1956). Rational choice and the structure of the environment. *Psychological Review* **63**: 129–138.

Smith, B. (1998). Basic concepts of formal ontolgy. In: N. Guarino (ed.). *Formal Ontology in Information Systems*. Amsterdam, Oxford, Tokyo, IOS Press: 19–28.

Timpf, S. (2002). Ontologies of wayfinding: a traveler's perspective. *Networks and Spatial Economics* **2**(1): 9–33.

Timpf, S., and A. U. Frank (1997). Metadaten – vom Datenfriedhof zur multimedialen Datenbank. *Nachrichten aus dem Karten- und Vermessungswesen Reihe* **I**(117): 115–123.

Timpf, S., M. Raubal, and W. Kuhn (1996). *Experiences with Metadata*. 7th International Symposium on Spatial Data Handling, SDH'96, Delft, The Netherlands (August 12–16, 1996), IGU.

Vckovski, A. (1997). *Interoperability and spatial information theory*. International Conference and Workshop on Interoperating Geographic Systems, Santa Barbara, CA (December 3–6, 1997).

Wand, Y., and R. Y. Wang (1996). Anchoring data quality dimensions in ontological foundations. *Communications of the ACM* **39**(11): 86–95.

Whitehead, A., and B. Russell (1910–1913). *Principia Mathematica*. Cambridge, Cambridge University Press.

A FLEXIBLE DECISION SUPPORT APPPROACH TO MODEL ILL-DEFINED KNOWLEDGE IN GIS

GLORIA BORDOGNA*, MARCO PAGANI
Consiglio Nazionale delle Ricerche, Istituto per la Dinamica dei Processi Ambientali, c/o POINT, via Pasubio 5, I-24040 Dalmine (BG), Italy

GABRIELLA PASI
Disco, Università di Milano Bicocca,via degli Arcimboldi 8, I-20128, Milano, Italy

Abstract. The contribution presents a flexible approach to model spatial decision strategies in GISs when either the data, or the knowledge of the phenomenon, or both, are affected by some form of imperfection. The proposal is particularly appealing for its flexibility and usefulness in many real applications encompassing geosciences, and environmental protection systems. Strategies based on fuzzy inference, soft integration of criteria with distinct importances, and consensual fusion are proposed and modeled within fuzzy set theory.

Keywords: flexible decision support, spatial data fusion, fuzzy inference, consensual fusion

1. Introduction

Current GISs are inadequate to support the experts in modeling decisions affected by uncertainty involving imperfect spatial information because flexible decision strategies can hardly be defined by using the available technologies (Burrough and Frank, 1996; Codd et al., 2000; Jankowski and Nyerges, 2001). Generally experts have to develop their ad hoc software tools to model a specific problem, such as to predict the behavior of a process, floods, landslide hazard, seismic hazard, water pollution distribution, impact factor

*To whom correspondence should be addressed. Gloria Bordogna, *Istituto per la Dinamica dei Processi Ambientali, Consiglio Nazionale delle Ricerche, c/o POINT, via Pasubio 5, I-24040 Dalmine(BG) – Italy; gloria.bordogna@idpa.cnr.it*

A. Morris and S. Kokhan (eds.), Geographic Uncertainty in Environmental Security, 133–152.

evaluation. These tools are often integrated within a GIS so as to exploit the spatial analysis functionalities offered by these systems (Malczewski, 2006). The criteria that they offer to support decisions are generally based on Boolean logic, basically maps overlay and weighted linear combination (Jiang and Eastman, 2000).

Many researchers in the field of GISs outlined the need of integrating flexible decision support functionalities to perform multicriteria evaluations to solve allocation or location problems, to perform suitability analysis, to integrate individual criteria for options choice and group decisions (Codd et al., 2000; Jankowski and Nyerges, 2001; Malczewski, 1999; Scott and Robinson, 2000). In this paper we propose a flexible and robust decision support approach that can be suited and personalized to define models of ill-known spatial phenomena.

The proposed approach is *flexible* in the sense that it allows the setting of distinct multicriteria decisions related to complex phenomena characterized by distinct levels of ill-defined knowledge. Specifically, it can manage:

- The definition and the evaluation of heuristic vague rules governing a phenomenon so as to produce new themes to support a decision purpose
- The reinforcement of partial evidences of a phenomenon by allowing the definition of soft integration strategies of spatial data so as to generate a map representing the global evidence of the phenomenon
- The definition of consensus fusion strategies to synthesize the results produced independently by groups of experts or by competitive models by taking into account their agreement so as to reduce possible semantic errors

The approach is *robust* in the sense that it copes with data of different type possibly affected by some kind of imperfection, such as measurement errors of the means of acquisition, approximation due to the adopted representation, incompleteness, etc.

Finally, the approach is founded on fuzzy logic (Zadeh, 1987) since it provides a unifying formal framework to represent and model both different states of ill-defined knowledge, and the imprecision/uncertainty of the available spatial data. Fuzzy sets support the representation of vague concepts such as imprecise, and uncertain values of attributes of spatial entities; soft aggregation operators allow to define distinct "soft" aggregation strategies of thematic layers that allow partial compensations among the themes (Jiang and Eastman, 2000; Yager, 2004); fuzzy rules and linguistic variables allow to directly code vague expert's knowledge. Finally fuzzy inference permits applying fuzzy rules to evaluate possibly imprecise data, thus producing results with an estimated degree of certainty (Zadeh, 1965).

Fuzzy logic has been successfully applied in spatial data analysis and multicriteria evaluation (Burrough and Frank, 1996; Morris and Jankowski,

2000, 2001; Robinson, 2003). Many papers have proposed and applied fuzzy logic to model spatial entities with ill-defined boundaries, to define flexible spatial query languages, and to define decision-making strategies (Bone et al., 2005; Bordogna et al., 2006; Brivio et al., 2006; Chanussot et al., 1999; Codd et al., 2000; Scott and Robinson, 2000; Solaiman, 1999; Tran et al., 2002).

The distinguishing characteristic of our proposal is to offer the expert the choice of alternative flexible decision strategies. According to the most appropriate representation of the expert knowledge, one can decide to describe the phenomenon in the form of fuzzy rules, or more simply in the form of integration of complementary hints of evidence of the phenomenon, or even by a consensual fusion of maps resulting from the application of competitive models.

In the first section we introduce the type of data that are managed. In section 3 the states of ill-defined knowledge that are represented and managed are exemplified. In section 4 the functionalities of the flexible decision approach are described and formalized. In section 5 an application example is discussed. Finally the results summarize the main achievements and future perspectives.

2. Characteristics of the Managed Spatial Data

In this section the kinds of managed spatial data are analyzed with respect to their "imperfection," intended as either imprecison/vaguess or uncertainty/indeterminacy.

GISs manage geographic entities that are represented by a set of properties, generally named themes, and a spatial reference, that is an attribute specifying the spatial location of the geographic entity with respect to a spatial domain.

In the context of the present contribution we consider the management of spatial data in raster form. Thus, prior to the application of the decision support approach, one needs to transform the vector data into grids so as to allow their processing.

As far as the pixel values are concerned, they can be specified with one among three types:

- *Numeric values:* for this type of data a metrics is defined and we can compute all types of arithmetic operations and a distance. Examples of such kind of data are the local slope or altitude of a spatial position; the density of some spatial property such as population, pollution, disease etc.
- *Ordinal values*: for these type of data an ordering is defined and some compositional operations based on the index of the labels can be defined, for example, a similarity or proximity relationship based on the dif-

ference between indexes. Examples of such data are the classes of hazard, risk, susceptivity etc.

- *Nominal values*: for this type of data the composition of values is meaningless; however, a similarity or proximity relationship for each pairs of values can be defined representing a property of the datum, e.g., a physical/chemical property. Examples of such data are the names of soil types, lithology types, etc. These values can be ordered with respect to some property, such as their favorability to contribute to the occurrence of a phenomenon (e.g., landslide).

As far as the pixel's spatial reference is concerned, it is generally represented by a pair of indexes (i,j) univocally identifying the pixel within the grid and the precise cell on the terrain. The geographic coordinates can be computed straightforward by knowing the pixels' resolution and the grid origin coordinates.

Imperfection in spatial data may affect either the pixel values or the pixels' spatial reference or both (Burrough and Frank, 1996; Fisher, 2000; Goodchild and Gopal, 1989). In the following, we analyze the representation within fuzzy set theory of the different kinds of imperfection of the spatial data.

2.1. IMPERFECTION OF NUMERIC VALUES

Numeric values can be imprecise and vague when they are not single elements of the numeric domain. This is the case of values obtained by statistic analysis expressing mean values of a property with an associated dispersion, such as daily temperatures.

Within fuzzy set theory, vague and imprecise numeric values can be represented by fuzzy subsets on the basic numeric domains. For example, vague values of local slope or altitude can be linguistically expressed by terms such as *low, medium, high* and represented by membership functions on the numeric values of slope [0°, 90°] and altitude [0,5000] respectively. Imprecise values can be represented by intervals of basic numeric values e.g., slope is [15°, 18°]: their membership function takes the maximum degree for all the elements in the interval and zero outside. Imprecise values can be used to represent cases of indeterminacy due to low resolution, or even to represent missing values.

In many real cases, the available data are crisp, precise, but one is uncertain on their reliability for several reasons: either because the agencies that is the source of the data cannot be completely trusted, or because one knows that the means of acquisition are not enough sophisticated and generate systematic errors; not least because data are a result of a subjective analysis, such as surveyed data. Uncertainty on precise or imprecise data can be represented by associating a degree of confidence or credibility, or reliability with them, e.g.,

- Slope is 30% with reliability r
- Slope is *high* with *very high* reliability

A compatibility relationship between imprecise/vague numeric values can be defined based on a *similarity* measure between fuzzy sets.

A common *similarity* relationship between fuzzy sets is the fuzzy Jaccard's coefficient. Given two fuzzy sets A and B defined on a discrete (continuous) domain D:

$$similarity(A,B) = \frac{\sum_{i=1}^{N} min(\mu_A, \mu_B)}{\sum_{i=1}^{N} max(\mu_A, \mu_B)} \qquad similarity(A,B) = \frac{\int_{i \in D} min(\mu_A, \mu_B)}{\int_{i \in D} max(\mu_A, \mu_B)} \tag{1}$$

In the case in which both A and B are imprecise $A = [a_{\min}, a_{\max}]$, $B = [b_{\min}, b_{\max}]$ their similarity measure can be defined as:

$$similarity(A,B) = \begin{cases} 0 & if\ a_{max} < b_{min}\ or\ b_{max} < a_{min} \\ \dfrac{min(a_{max}, b_{max}) - max(a_{min}, b_{min})}{max(a_{max}, b_{max}) - min(a_{min}, b_{min})} & otherwise \end{cases} \tag{2}$$

2.2. IMPERFECTION OF ORDINAL VALUES

Ordinal values can be imprecise and vague when one is unable to specify a single value of the ordinal domain but can select a set of values e.g.,

- A point of a map may be labeled as both *high* or *very high* risk.

Uncertainty can be specified by associating with an ordinal value a reliability degree, e.g.,

- A point of a map may be labeled as *high* with *r* reliability.
- A point of a map may be labeled as *high* with *very high* reliability.

A compatibility relationship can be defined for imprecise ordinal values based on a similarity relationship.

A simple definition of the similarity between two sets of labels A and B can be defined by considering the indexes of their elements on the ordinal scale, and by applying definition (2) in which a_{min}, a_{max} and b_{min}, b_{max} are the minimum and the maximum indexes of the labels in A and B respectively.

2.3. IMPERFECTION OF NOMINAL VALUES

Ambiguous categorizations derived by the inability to associate a single nominal value with a spatial position arise the need to define imprecise nominal values. For example, this occurs in many applications of remote sensing when one has to classify a region into a soil type based on the analysis of its spectral signature. In these situations it can be useful to associate several soil types with the same spatial position thus defining a mixture type element. These mixture values can be represented by fuzzy sets on the discrete basic domain of the original nominal types.

A similarity or proximity degree between imprecise nominal values can be evaluated in the features space representing the nominal values; for example, this occurs when applying clustering techniques. A nominal value can be associated with a confidence degree or label to a spatial element representing this way the reliability of the value.

2.4. IMPERFECTION OF THE PIXEL SPATIAL REFERENCE

Imperfection may also affect the spatial reference of the pixel, not just its value. Indeterminacy is inevitably introduced for example when rescaling an image to match a given resolution required by the analysis (Burrough and Frank, 1996; Fisher, 2000; Goodchild and Gopal, 1989).

To take account of the spatial indeterminacy of a pixel one can represent its spatial reference, usually identified by (i,j), through a fuzzy relation $R_{i,j}$:X×Y→[0,1] on the bidimensional spatial domain X×Y, e.g., defined by the Cartesian product of two trapezoidal membership functions $(a_i, b_i, c_i, d_i) \times (a'_j, b'_j, c'_j, d'_j)$. This way the grid becomes a fuzzy grid (Schneider, 2003).

Imprecision in the pixels' spatial reference sometimes can be inversely related to the imprecision in pixels' values. For example, when one has to map the border between two fuzzy ecosystems one can either choose to associate a unique class with the pixel value and an imprecise spatial reference, or can define a precise spatial reference and several soil classes as imprecise value of the pixel. If one has to evaluate the environmental impact of a new road in a region that is traced on the border of two different fuzzy ecosystems characterized by different impact factors indicators it is important to be able to take into account that the border between these ecosystem is fuzzy by very nature (Bordogna et al., 2006). By representing the ecosystems with an imprecise spatial reference one can weigh the contributions of the crisp pixel values depending on their possibility degree to represent a given position on the ground.

3. States of Ill-Defined Knowledge

In this section the imperfection of the expert's knowledge relative to an environmental phenomenon is analyzed and its representation is proposed within fuzzy set theory.

In many real cases decisions on a territory such as those involved in allocation/location, suitability assessment, impact assessment, require spatial analysis operations to generate higher quality maps, named decision maps, carrying a specific semantics (Malczewski, 1999; Valet et al., 2001). Examples of such maps are the hazard/vulnerability/risk maps of a land to an environmental phenomenon. To automatically generate decision maps a model of the phenomenon under study is needed, or, when a model is not available, at least historical data on the occurrence of the event on the considered land are necessary in order to apply learning techniques based on training (Malczewski, 1999; Valet et al., 2001). Nevertheless, historical data are rarely available, and since environmental phenomena are often complex, the generation of decision maps has to rely on the knowledge available, often incomplete and vague (Brivio et al., 2006; Malczewski, 1999; Valet et al., 2001).

Three levels of "imperfection" of the knowledge of phenomena are considered:

1. The expert can express the laws governing the phenomenon in the form of vague rules. He/she can select a set of variables influencing the phenomenon, and can state the vague/imprecise critical values of the variables that alone or in combination one another may determine the output variable values, that is the values of the decision map.
2. The expert can only identify a set of vague and complementary hints of partial evidence of the phenomenon, and the decision map is modeled to reflect a global evidence of the phenomenon. Differently than in the previous situation, he/she cannot specify interactions among the critical values of the variables, and is unable to quantify precisely the minimum number of hints that are necessary for stating with certainty the occurrence of the phenomenon. In this case the overall global evidence is reinforced by an accumulation of hints (the more the hints, the more the global evidence).
3. Several decision maps are available, generated by competitive models or experts, each one characterized by a distinct reliability (credibility), and we want to fuse their possibly contradictory values so as to achieve a more reliable consensual decision map.

Let us analyze how these states of knowledge can be represented within fuzzy set theory.

3.1. 1ST STATE OF KNOWLEDGE

The expert's knowledge of the phenomenon can be represented in the form of fuzzy rules (Zadeh, 1987):

$$If\ A_1\ op,...,op\ A_N\ then\ D$$

in which A_i $I = 1,...,N$ are fuzzy predicates such as *high, medium, low*, that is, linguistic values in the term set of the phenomenon variables and D is a vague value of the output variable. A defuzzification of the output values is sometimes needed to generate a numeric decision map. The aggregation operators *op* in the antecedent of the rules can be logical connectors *and, or, not* or a fuzzy aggregation operator such as the OWA operator (Yager, 1988) expressed by a linguistic quantifier like *most of.* These operators specify the interaction between the variable critical values.

The phenomenon is modeled by applying the fuzzy rules through a fuzzy inference process on a set of spatial data layers associated with the input variables. Fuzzy inference can be applied even in the case in which the data are affected by imprecision both in their values and spatial reference.

Example: *Landslide susceptivity maps*

A decision map expressing the susceptivity of a land to landslides can be generated by evaluating a set of spatial data layers based on a set of fuzzy rules describing the landslide phenomenon. Influencing input variables are identified as the slope, lithology, soil usage, internal relief, with linguistic values on a basic domain of the variables (e.g., slope has a continuous basic domain [0°, 90°], while lithology has a nominal domain $\{a,b,c,d,e,f\}$). The output susceptivity degree is defined with linguistic values such as: *none, low, medium high, full*. Examples of fuzzy rules describing the susceptivity to landslide can be found in Guerrieri (2006); they look like the following ones:

If slope is high and lithology is sand then susceptivity is very high
If slope is medium and soil usage is not cultivate then susceptivity is high
If slope is high and lithology is mixed then susceptivity is medium

Slope, lithology, etc. are associated with distinct spatial data layers taking values in a basic domain of the variables.

The susceptivity map is generated by the application of the fuzzy rules to the data layers associated with the variables.

If all the data layers have the same resolution and spatial reference system of the output map, the rules are iteratively applied pixel by pixel independently.

In the other cases, first we have to transform each input map to the same resolution of the output map. Notice that in the transformation we may generate imprecise values of the pixels.

3.2. 2ND STATE OF KNOWLEDGE

The phenomenon is modeled using a multisource spatial data integration operation.

The global evidence of a phenomenon that must be estimated and represented in the decision map D is determined through a reinforcement of compensatory hints (partial evidences) so that *the more the hints the more the global evidence*. The utility of fuzzy aggregation operators to model this kind of knowledge has been proposed and applied in various contexts (Bone et al., 2005; Brivio et al., 2006; Chanussot et al., 1999; Jiang and Eastman, 2000; Robinson, 2003).

The knowledge of the phenomenon is expressed through N soft constraint C_i, $i = 1,..M$, expressed by linguistic terms and defined by fuzzy subsets on the domains of the variables associated with input data layers and aggregated by a linguistic quantifier Q:

$$Q(C_1, ..., C_M) = D$$

The soft constraints specify the (vague) critical or anomalous values of the variables that may hint to the phenomenon. The degrees of satisfaction of the soft constraints by the input data layers are interpreted as strengths of the hints.

The relative quantifier Q, defined by a monotone not decreasing fuzzy set $\mu_Q:[0,1]\rightarrow[0,1]$, and expressing a fuzzy majority, specifies the kind of soft integration strategy (Yager, 2004; Yager, 1996). It indicates what kind of compensation between the hints must be applied to compute the degree of overall global evidence, i.e., Q = *at least 1* full compensation, Q = *all* no compensation, Q = *most of* partial compensation.

Sometimes importance weights i_{C1} ..., i_{CM} (ordinal or numeric weights) can be associated with the soft constraints $C_1, ..., C_M$ to indicate their role in determining the global evidence.

The data layers associated with the constrained variables are evaluated to compute their degrees of satisfaction of the soft constraints. These degrees are interpreted as partial evidence values of the phenomenon.

The integration operation is applied independently for each pixel in the case where the resolution of the output map is the same of the input maps. In the other cases, first we have to transform each input map to the same resolution of the output map, and then apply the integration operation. Notice that in the transformation we may generate imprecise values of the pixels, so the integration operation must be defined to cope with imprecise values.

TABLE 1. Soft constraints on variables

Soft constraint	Variable name
Long	*Period of low precipitations below season's average*
Long	*Period of high temperature above season's average*
Low	*Reserves*
High	*Consumption for agriculture*
High	*Urban consumption*
High	*Industrial consumption*
High	*Distance from water basins*
Low	*Water quality*
Bad	*Usage policy*

Example: *Water shortage*

The estimation of the global evidence of water shortage on a land can be described by the occurrence of a fuzzy majority of hints whose strength is computed by the evaluation of following soft constraints on the data layers representing the variables in Table 1:

The soft constraints admit degrees of satisfaction in [0,1] thus making comparable the strengths of the hints to integrate (Jiang and Eastman, 2000). Some variables such as *period of Low precipitations below season's average* are generally not directly available but are a result of other decision processes (Brivio et al., 2006).

3.3. 3RD STATE OF KNOWLEDGE

There are distinct competitive models (e.g., corresponding to experts, judges, software tools) of the phenomenon each one with its own reliability (credibility). Each model generates its own decision map that can be affected by imprecision. It can be useful to generate a consensual decision map (Jankowski and Nyerges, 2001).

We model the global decision map based on a fuzzy consensual fusion strategy that aggregates the possibly contradictory imprecise maps generated by the competitive models taking into account their compatibility and reliability based on the following criteria:

- The more reliable is the model (expert), the more accurate and precise are considered the values of the pixels in the map generated by the model.
- The higher the agreement of a model with the others, the more the model contributes the consensual decision map.

This state of knowledge is represented by the reliability scores $r_1 \ldots r_n \in [0,1]$ of the models and the fusion criterion specified by a linguistic quantifier Q modeling the decision attitude.

Q = *all* means that the pixel values in the consensual map must reflect the common decision of all the models: in the case in which the values in the input maps are proportional to an alarm or anomaly condition, by specifying *all* one wants to model a secure attitude, that is, all the experts must agree on the need to issue the alarm or to point at the anomaly; this is useful to model a safe decision.

Q = *at least 1* means that the pixel values in the consensual map must reflect the highest value among those of the models; in the case in which the value is proportional to an alarm or anomaly, by selecting *at least 1.* one models a confident attitude, that is one trusts the most alarming model. This is useful in making precautionary decisions.

Q = *most of* means that the pixel values in the consensual map must reflect the shared decision of a fuzzy majority of models; this models a trade-off decision.

Example: *Evaluation of a consensual seismic hazard map*

The expert generates eight seismic hazard maps with the same resolution by applying distinct competitive models. Each model has a distinct reliability as defined in the literature (Rabinowitz et al., 1998).

A consensual seismic hazard map can be generated by applying a consensual fusion of a fuzzy majority of the most reliable models.

4. The Flexible Decision Support Approach

In this section the proposed decision support functionalities are presented and formalized (Figure 1).

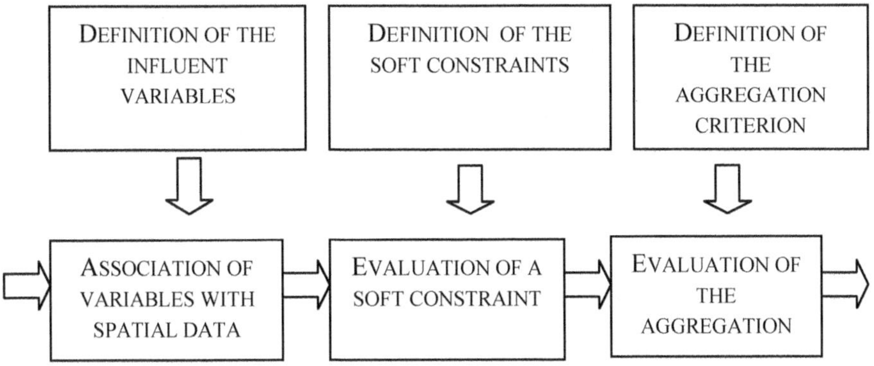

Figure 1. Functionalities of the Decision Support Approach

4.1. DEFINITION OF THE INFLUENT VARIABLES

The variable influencying the phenomenon are represented by a triple:

$$(vname,\ type\text{-}domain,\ basic\text{-}domain)$$

in which:

- *Vname* is the name of the variable.
- *Type-domain* specifies if the domain is numerical, ordinal, or nominal.
- *Basic-domain* is the domain of values and can be:
  A range [a, b] for numeric type domains
  A set of ordered labels $\{o_1,...o_M\}$for ordinal domains
  A set of labels $\{l_1, ...l_M\}$for nominal domains

4.2. DEFINITION OF THE SOFT CONSTRAINTS

Soft constraints can be defined as fuzzy sets on the domain *basic-domain* of the variables *vname*.

A flexible way to define increasing, decreasing, and unimodal soft constraints is by means of trapezoidal membership functions on continuous type domains, and by discrete fuzzy sets for discrete domains (see Figure 2). Soft constraints can be defined to represent the semantics of the linguistic values of the variables used to express fuzzy predicates in the rules. In this case the degree of satisfaction of the soft constraint is interpreted as the degree of compatibility of the datum with the fuzzy predicate.

If the soft constraint defines the critical values of a variable contributing to the phenomenon, its satisfaction degree is interpreted as the strength of the hint. The evaluation of a soft constraint serves also to normalize in the same domain [0,1] the values to integrate thus achieving their comparability.

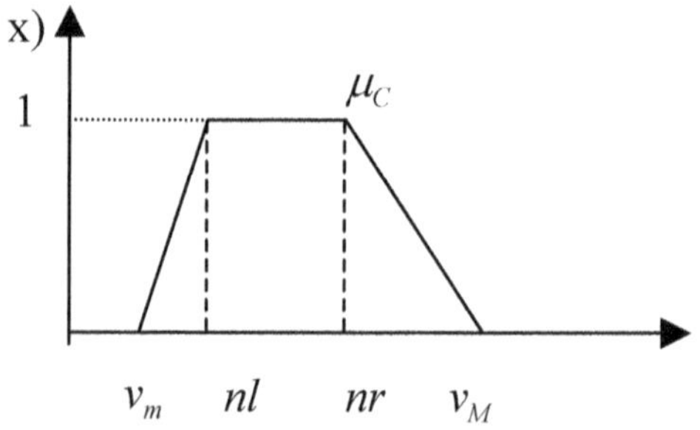

Figure 2. Membership function of a soft constraint *C*

A soft constraint is defined by a triple:

$$(C,\ vname,\ \mu_C)$$

in which

- C is the name of the constraint.
- *Vname* is the name of the constrained variable.
- μ_C specifies the semantics of the constraint, that is, its membership function. μ_C is formalized depending on the *type-domain* of the variable *vname*.

For continuous *type-domain:*

$$\mu_C = (v_m\ n_m\ n_M\ v_M)$$

with $v_m\ n_m\ n_M\ v_M \in$ *basic-domain* of *vname*. It identifies a trapezoidal membership function.

For discrete *basic-domain*:

$$\mu_C = \{m_1/v_1\ ...,\ m_M/v_M\}$$

with $m_i \in [0,1]$ and $v_i \in$ *basic-domain* of *vname* that can be ordinal and nominal *type-domain.*

4.3. DEFINITION OF THE AGGREGATION CRITERION

The aggregation criterion can be of three types depending on state of ill-defined knowledge that has to be represented.

4.3.1. *Fuzzy rules*

In the case of the first state of knowledge the aggregation criterion specifies the set of fuzzy rules and the meta information for the fuzzy inference.
A fuzzy rule is represented by a pair (*antecedent, consequent*): in which *antecedent* and *consequent* are well-defined strings recognized in the language:

antecedent: = *<fuzzy predicate>* {, *<op>* *<fuzzy predicate>*} |
<quantifier> {N,} *<fuzzy predicate>*
consequent: = *<fuzzy predicate>*
<fuzzy predicate>: = *vname* "is" *C*
<op>: = *and* | *or* | *and not*
<quantifier>.= *all, at least N, most of*

and C is a soft constraint name

The antecedent is defined in terms of the fuzzy predicates defined by soft constraints C on the variables *vname* combined by means of the operators *and, or, not* and *quantifier*. The meta information associated with the execution of the fuzzy inference rule is defined by the tuple:

(*Implication, composition, defuz; and-def, or-def, not-def, Q*);

This tuple specifies the names of the functions defining the kind of implication, composition, and defuzzification of the fuzzy rules, and the functions associated with the aggregation operators *and, or, not*, and *quantifier* used in the rules, e.g. (*Larsen, max-min, Center-of-gravity, min, max, 1-complement, most of*). $\mu_{mos\text{-}of}$ $[0,1]\rightarrow[0,1]$ is the membership function formalizing the semantics of the quantifier *most of* as a fuzzy set.

4.3.2. *Integration of hints*

In the case of the second state of knowledge the aggregation criterion corresponds to a soft integration function of compensatory values possibly with distinct importance degrees. The aggregation criterion is represented by a triple:

$$(\underline{Q}, \underline{C}, \underline{I})$$

in which Q is a linguistic quantifier name such as *most of, averagely all*, etc. identifying a fuzzy set $\mu_Q : [0,1]\rightarrow[0,1]$; a simple representation of a monotone not decreasing definition of μ_Q is provided by specifying a triple (a, b, c) (see Figure 3) in which a is value where the quantifier Q starts to increment, b where it starts to give full satisfaction degree and $c > 1$ the smoothing factor of the interval [a, b]

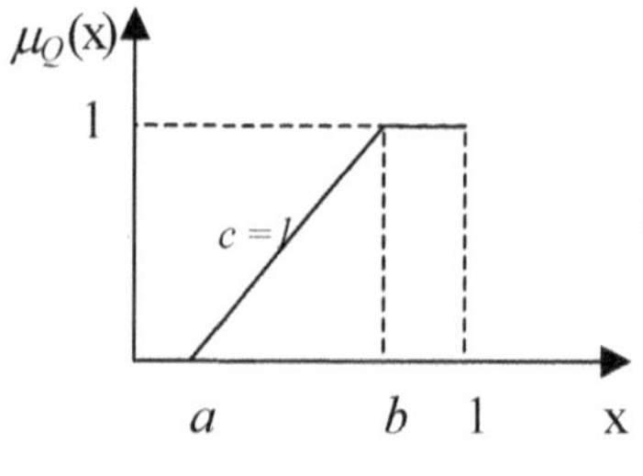

$$\mu_Q(x) = \begin{cases} 0 & x \leq a \\ \dfrac{x^c}{b-a} - \dfrac{a}{a-b} & a < x < b \\ 1 & x \geq b \end{cases}$$

Figure 3. Membership function of a linguistic quantifier Q

$\underline{C}$ is an n-dimensional vector of soft constraints names and $\underline{I}$ is an n-dimension vector of importance degrees in [0,1] associated with the soft constraints and representing their influence in the integration.

4.3.3. *Consensual fusion*

In the case of the third state of knowledge the aggregation criterion identifies a consensual fusion function of compatible values with distinct reliability. It is represented by a triple:

$$(Q, \underline{D}\text{-}Layers, \underline{R})$$

in which Q is a the name of a linguistic quantifier μ_Q identifying a fuzzy majority; $\underline{D}$*-layers* is a vector of n names of *d-layers* univocally identifying data layers, and $\underline{R}$ is a vector of n reliability degrees in [0,1] associated with the n *d-layers*.

4.4. ASSOCIATION OF VARIABLES WITH SPATIAL DATA LAYERS

This operation consists in associating a data layer, represented by a raster map *d-layer* with resolution *rel*, with a variable *vname*. The consistency of the domain of the spatial data layer with the *basic-domain* of the specified variable has to be checked. The values of the pixels in *d-layer*, can be imprecise values on *basic-domain*, as defined in section 2.

In many real cases it is often necessary to generate a data layer by applying spatial operations to other data available. Here we assume that these operations can be performed by exploiting the functionalities commonly available in GISs.

4.5. EVALUATION OF THE SOFT CONSTRAINTS

The evaluation of soft constraints serves a twofold purpose:

- To compute the degrees of satisfaction of the soft constraint by the data, that is interpreted either as strength of the hints of the phenomenon or as degrees of compatibility of fuzzy predicates in the rules.
- To normalize the values of strength of the hints to the same domain thus achieving their comparability necessary for the subsequent integration.

In the case in which the data layers have imprecise values, degrees of strength of the hints are computed by applying a compatibility relationship function defined for the type of data (see section 2 definitions (1) and (2)).

4.6. EVALUATION OF THE AGGREGATION FUNCTION

This functionality allows three kinds of aggregations.

4.6.1. *Fuzzy inference*

When a set of fuzzy rules is provided the aggregation operation consists in their application to the selected data layers, possibly imprecise or vague, by the evaluation of the fuzzy inference rule (Generalized Modus Ponens) (Zadeh, 1987).

4.6.2. *Integration of degrees of strength of the hints*

In this case the aggregation operation consists in the integration of the strengths t_1 , ..., $t_n \in [0,1]$ of the hints by weighting their contribution with their importances i_1 , ..., i_n. To this end the integration function is defined by an OWA operator of dimension N associated with the linguistic quantifer Q (Yager, 1988; Yager, 1996):

$$\mathrm{OWA_Q}(t_1, ..., t_n) = \sum_{\mathrm{i = 1,...n}} w_i * b_i \quad \text{with } \sum_{\mathrm{i = 1,...n}} w_i = 1$$

and b_i the i-th greatest of the t_1 , ..., t_n degrees of satisfaction of the soft constraints by the same spatial element (pixel).

From the definition of the quantifer μ_Q : [0,1] → [0,1] the weighting vector W of the OWA_Q operator is derived by computing the following:

$$w_i = \mu_Q\left(\frac{1}{e}\sum_{j=1}^{i} e_j\right) - \mu_Q\left(\frac{1}{e}\sum_{j=0}^{i-1} e_j\right) \qquad e = \sum_{i=1}^{K} e_i = \sum_{j=1}^{K} i_j$$

where e_j is the importance degree associated with the j-th largest t_j. This way, the increment in satisfaction in having i non-null values with respect to i-1, that is, w_i increases with the importance e_i. The values having no importance play no role.

4.6.3. *Consensual fusion of competitive models*

In this case the aggregation function consists in a consensual fusion correspondent to a fuzzy majority expressed by a monotone nondecreasing linguistic quantifier Q of possibly imprecise values v_1, ..., v_n with associated distinct reliability degrees r_1 , ..., r_n.

The definition of the weighting vector W of the OWA_Q operator is obtained as in the previous case from μ_Q.

The importance degree i_i of a value v_i to fuse expresses the degree of agreement of the value v_i with respect to the other values v_k with $k \neq i$ weighted by its associated reliability degree r_i. In this way, the contributions of the models with highest agreement with the other models and highest reliability are more heavily taken into account in the fusion.

The degree of importance i_i is computed by applying the following formula based on a compatibility relationship defined depending on the type of values (either numeric, ordinal, nominal, precise, or imprecise) as discussed in section 2:

$$i_i = r_i \frac{\sum_{k=1,k\neq i}^{n} compatibility(v_i, v_k)}{\max_{i=1..n}\left(\sum_{k=1,k\neq i}^{n} compatibility(v_i, v_k)\right)}$$

In the case in which the values to fuse are numeric and precise values the classic OWA definition given previously is used. In case of ordinal values the OWA operator can be applied to the indexes of the ordinal values.

For imprecise values, i.e., intervals $[v_{1m}, v_{1M}],\dots,[v_{nm}, v_{nM}]$, one can apply the consensual fusion to intervals by extending the OWA definition with fuzzy arithmetic operations as follows:

$$OWA_Q([v_{1m}, v_{1M}],\dots,[v_{nm}, v_{nM}]) = \sum_{i=1,\dots n} w_i * [b_{im}, b_{iM}]$$
$$\text{with } \sum_{i=1,\dots n} w_i = 1$$

and $[b_{im}, b_{iM}]$ is the ith greatest of the $[v_{1m}, v_{1M}],\dots,[v_{nm}, v_{nM}]$ such that:

Order:	$[a_1, a_2] > [b_1, b_2]$ if $(a_1 + b_1) / 2 > (a_2 + b_2) / 2$
Addition:	$[a_1, a_2] + [b1, b2] = [a_1 + b_1, a_2 + b2]$
Product:	$[a_1, a_2] \times [b1, b2] = [a_1 \times b1, a2 \times b2]$

In the case in which the values to fuse are vague values represented by a possibility distribution μ_v the consensual fusion is applied to their $\tilde{r}_i$-cuts(μ_v), i.e., to intervals.

$\tilde{r}_i$-cuts(μ_v) is obtained by considering the reliability degree r_i. This way, the more reliable (certain) values are interpreted as bearing less imprecision.

5. Examples of Application: Generation of a Consensual Seismic Hazard Map

As an example of application of the consensual fusion strategy the generation of a consensual seismic ground motion map is described.

Eight competitive seismic models of ground motion are applied to the classification of the same territory, each one with a reliability score.

In the classical approach the fused map is generated as the weighted average of the maps in which the weight is the reliability degree (Rabinowitz et al., 1998).

In our flexible approach, we take into account the imprecision of the models in generating the maps by representing the ground motion value of a pixel through a fuzzy number (g_l, g, g_r) in which g is the ground motion value computed by the model in the current pixel, and g_l and g_r are defined to capture the imprecision of the computation.

A consensual fusion strategy is applied to generate the consensual ground motion map relative to a fuzzy majority of the reliable models.

We want to model a trade-off decision strategy.

The fuzzy majority *most of* is represented by the fuzzy set $\mu_{most\ of}$ with non decreasing membership function ($a = 0.25$, $b = 0.75$, $c = 1$).

The maps produced by the eight models are simulated so that their ground motion values have a Gaussian distribution with distinct centers. In Figure 4 we have a representation of the sources of the Gaussian distributions produced by the eight models. Notice that six of the sources are close to each other, while sources 3 and 4 are quite far away.

In Figure 5A we have a map generated by the classic approach based on the weighted mean. The gray level is proportional to the ground motion value. In Figure 5B we depicted the map generated with the consensual fusion based on the OWA. The comparison of the two figures evidences how the consensual approach reinforces the ground motion values of the pixels in the regions where the centers of the Gaussian distributions are close each other, while it decreases the ground motion values of pixels where there is disagreement among the models with respect to the classic approach.

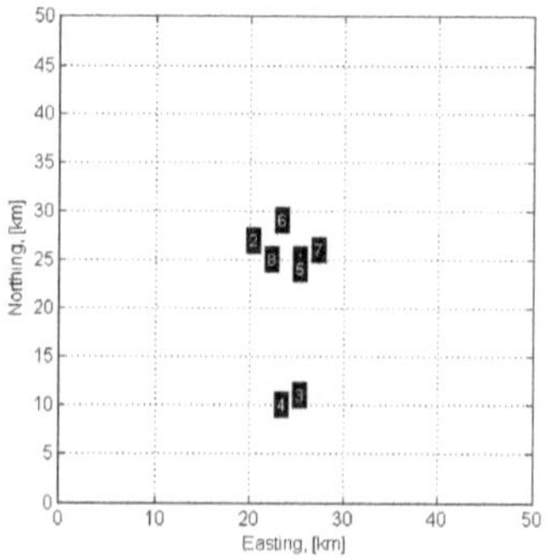

Figure 4. Centers of the Gaussian distributions simulating the spatial distributions of ground motion

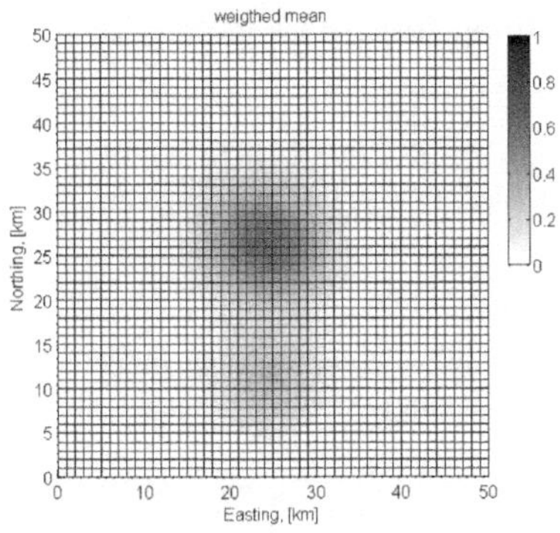

Figure 5A. Ground motion map obtained via the classic weighted mean aggregation

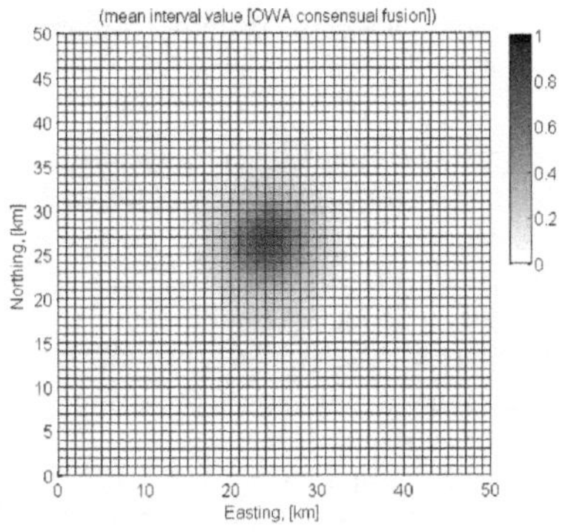

Figure 5B. Ground motion map obtained via the consensual OWA

6. Conclusions

In this constribution we have proposed some flexible decision strategies based on the representation and management of both imperfect spatial data and ill-defined knowledge in the context of fuzzy set theory. We have discussed how fuzzy sets can provide means to represent distinct state of ill-defined knowledge and distinct kinds of data imperfection. Some approaches such as the representation of the knowdge of phenomena through fuzzy rules and the use of integration strategies based on fuzzy measures have been already proposed in the context of GISs. We proposed a novel consensual fusion strategy that allows to aggregate both imprecise values with distinct reliability by taking into account their agreement.

References

Bone, C., Dragicevic, S., and Roberts, A. (2005), Integrating high resolution remote sensing, GIS and fuzzy set theory for identifying susceptibility areas of forest insect infestations. International Journal of Remote Sensing 26(21), 4809–4828.

Bordogna, G., Chiesa, S., and Geneletti, D. (2006), Linguistic modelling of imperfect spatial information as a basis for simplifying spatial analysis, Information Sciences, Elsevier, 176, 366–389.

Brivio, P.A., Boschetti, M., Carrara, P., Stroppiana, D., and Bordogna, G. (2006), Fuzzy integration of satellite data for detecting environmental anomalies across Africa. In J. Hill and A. Roeder (eds), Advances in Remote Sensing and Geoinformation Processing for Land Degradation Assessment, London, Taylor & Francis.

Burrough, P.A. and Frank, A.U., eds. (1996), Geographic Objects with Indeterminate Boundaries, GISDATA series, Taylor & Francis.

Chanussot, J., Mauris, G., and Lambert, P. (1999), Fuzzy fusion techniques for linear features detection in multitemporal sar images. IEEE Transactions on Geoscience and Remote Sensing 37(3), 1292–1305.

Codd, M., Petry, F., and Robinson, V., eds. (2000), Uncertainty in Geographic Information Systems and Spatial data, special issue of Fuzzy Sets and Systems, 113(1), 1–159.

Fisher, P. (2000), Sorites paradox and vague geographies, Fuzzy Sets and Systems, 112(1), 7–18.

Goodchild, M.F. and Gopal, S., eds. (1989), Accuracy of Spatial Databases, Taylor & Francis, London.

Guerrieri, V. (2006), Integrazione in un GIS di un modulo flessibile per la modellazione della suscettività alle frane e applicazione a un caso di studio concreto, thesis University Milano Bicocca.

Jankowski, P. and Nyerges, T. (2001), Geographic Information Systems for Group Decision Making, London, Taylor & Francis.

Jiang, H. and Eastman, J.R. (2000), Application of fuzzy measures in multi-criteria evaluation in GIS, International Journal of Geographical Information Science, 14(2), 173–184.

Malczewski, J. (1999), GIS and Multicriteria Decision Analysis, New York, Wiley.

Malczewski, J. (2006), GIS-based multicriteria decision analysis: a survey of the literature, International Journal of Geographical Information Science, 20(7), 703–726.

Morris, A. and Jankowski, P. (2000), Combining Fuzzy Sets and Databases in Multiple Criteria Spatial Decision Making. FQAS 2000, 103–116.

Morris, A. and Jankowski, P. (2001), Fuzzy techniques for multiple criteria decision making in GIS. IFSA World Congress and 20th NAFIPS International Conference, 2001. Joint 9th.

Rabinowitz, N., Steinberg, D.M., and Leonard G. (1998), Logic trees, sensitivity analysis and data reduction in probabilistic seismic hazard assessment, Earthquake spectra, 14(1), 189–201.

Robinson, V.B. (2003), A perspective on the fundamentals of fuzzy sets and their use in geographic information systems. Transactions in GIS 7(1), 3–30.

Schneider, M. (2003), Design and implementation of finite resolution crisp and fuzzy spatial objects, Data and Knowledge Engineering, 44, 81–108.

Scott M.D. and Robinson, V.B. (2000), A multiple criteria decision support system for testing integrated environmental models, Fuzzy Sets and Systems, 113, 53–67.

Solaiman, B. (1999), Multisensor data fusion using fuzzy concepts: application to land-cover classification using ERS-1/JERS-1 SAR composites. IEEE Transactions on Geoscience and Remote Sensing, 37(3), 1316–1326.

Tran, L.T., Knight, C.G., O'Neill, R.V., Smith, E.R., Riitters, K.H., and Wickham, J. (2002), Environmental assessment, fuzzy decision analysis of integrated environmental vulnerability assessment of the Mid-Atlantic region. Environmental Monitoring 29(6), 845–859.

Valet, L., Mauris, G., and Bolon, P. (2001), A statistical overview of recent literature in information fusion. IEEE AESS Systems Magazine 7–13.

Yager, R.R. (2004), A framework for multi-source data fusion, Information Sciences 175–200.

Yager, R.R. (1988), On ordered weighted averaging aggregation operators in multi-criteria decision making. IEEE Transactions on Systems, Man and Cybernetics 18, 183–190.

Yager, R.R. (1996), Quantifier guided aggregation using OWA operators. International Journal of Intelligent Systems 11, 49–73.

Zadeh, L.A. (1987), The concept of a linguistic variable and its application to approximate reasoning, parts I, II. Information Science, 8, 199–249, 301–357.

Zadeh, L.A. (1965), Fuzzy sets, Information and Control, 8, 338–353.

DEVELOPMENT OF THE GEOINFORMATION SYSTEM OF THE STATE ECOLOGICAL MONITORING

VITALIY B. MOKIN*
Vinnytsia National Technical University, Khmelnytske Shose 95, 21021, Vinnytsia, Ukraine

Abstract. The paper presents the characteristics of the created and introduced GIS on the level of the city, oblast, country, and the basin of the river, which flows in seven oblasts. These systems allow to solve different tasks on simulations and prognostication of changing and controlling over the ecological situations, including the application of the fuzzy sets theory.

Keywords: GIS, environment state monitoring systems, Vinnytsia region, Ukraine, water quality evaluation, surface water monitoring, hydrometeorology monitoring, observation station, fuzzy sets theory, Web-portal monitoring systems, automated systems, Southern Booh

1. GIS of the State Environmental Monitoring for Regional (oblast) Level

1.1. ENVIRONMENTAL PASSPORT OF THE VINNYTSIA REGION

We have achieved much on the level of the oblast. Now we are conducting works on the creation of the Environmental Passport of the Vinnytsia region. The customer of these projects is the State Department of Environment and Natural Resources Management in Vinnytsia Region of the Ministry for Environmental Protection of Ukraine. The performer is my Scientific and Research Laboratory of Ecological Development and Ecological Monitoring in Vinnytsia National Technical University.

The largest results are achieved in development of the subsystem of surface water monitoring in Vinnytsia Region.

* To whom correspondence should be addressed. Vitaliy B. Mokin, Department of the Simulation and Monitoring of Complicated Systems, Vinnytsia National Technical University, Khmelnytske shose 95, 21021, Vinnytsia, Ukraine; e-mail: vmokin@vstu.vinnica.ua

A. Morris and S. Kokhan (eds.), Geographic Uncertainty in Environmental Security, 153–165.

1.1.1. *Subsystem of surface water monitoring*

One of the main elements of computer monitoring system is the data bank of environmental information. It has almost 60 tables and 50 forms, special main button form, and a tool bar.

The data bank management system includes (Mokin, 2005):

- A form for analysis of overcoming data of boundary allowed values
- A form for finding minimum, maximum, and average values for many criteria
- A form for selecting data from bank for the subsequent construction of thematic maps
- Forms for working with passport of rivers, reservoirs, and ponds
- A form on reviewing the waste water discharges and water intakes

The main element of computer monitoring system is program shell, which combines the data bank and electronic map in the geoinformation system. This project uses Russian geoinformation program package "Map 2000"/"Map 2005" of Panorama Group (http://www.gisinfo.ru).

GIS Panorama has the set of powerful instruments for work with features and surface qualities of the objects. There is also the advanced interface allowing to create the user program modules. There is the multi-functional demo version of the package. The system ensures the optimal application of the computer recourses for dynamic drawing of the graphic images. There are the versions for Windows and Linux. Also there are the Ukrainian and English versions. It allows for different operations to work with the digital models of relief: 3D visualization, profile drawing, composing of the flood zone. It also allows to get the 3D relief picture of the flooded area. It ensures perfect relief visualization by regulation of the shadow of the relief forms.

Now, let us switch over to our soft. Program shell works with maps of territory of Vinnytsia region and of large-scale map of cities and other provinces of its region (Figure 1). There is the Ukrainian and English version of the program interface.

It enables to select in shell the object on the map and to obtain its information from data bank, for example (Mokin, 2005):

- The water user is selected – information on its water intakes and its water discharges
- The river is selected – hydrographical river parameters, chart of water quality along the river, parameters for riverbed, map, and photos
- The reservoirs and ponds are selected – main parameters and data

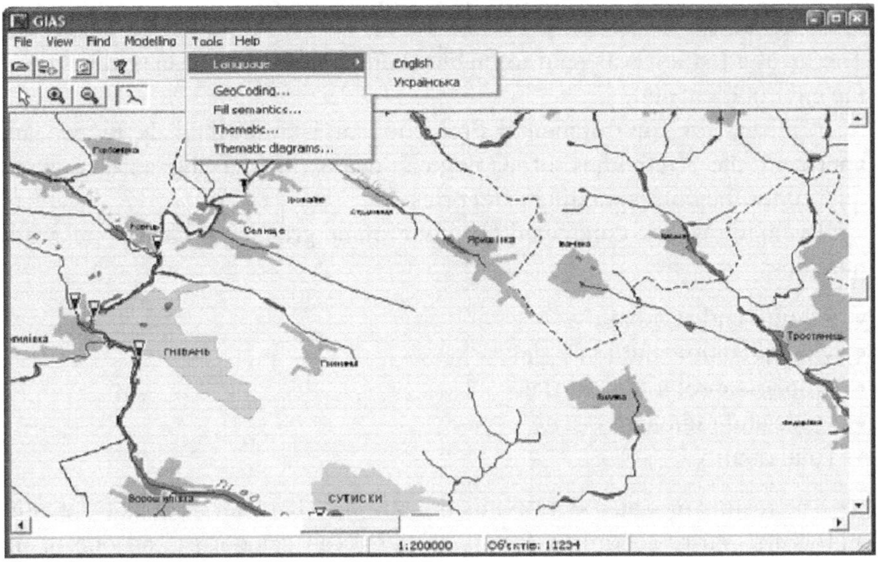

Figure 1. Program shell of computer monitoring system with map of territory of Vinnytsia region

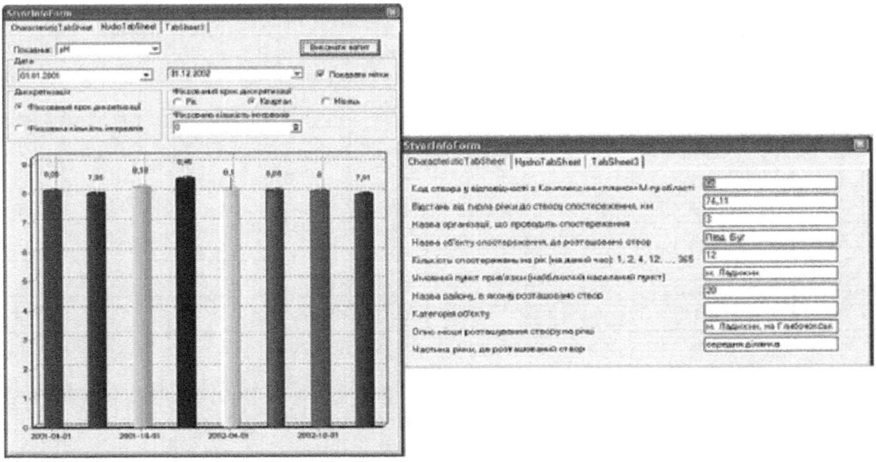

Figure 2. Information about of water quality observation station ("Stvor")

Besides, it enables to select in shell and to obtain information from data bank about water quality observation station. It is general information and chart of water quality according to the given period and given step (Figure 2).

This GIS is used for simulation of the change water quality and for solving many different tasks (Mokin, 2003).

1.1.2. *Methods and approaches of classification of the environment state*

The great attention was paid to methods and approaches of classification of the environment state.

Ministry for Environmental Protection of Ukraine had developed and approved the Methodics of ecological evaluation of the surface water quality for the corresponding categories.

Evaluations are conducted for four main groups of factors of water qualities:

- Salinity and salt state
- Mineralization and ionic state
- Tropho-saprobiological state
- Toxic and radioactive state
- Total quality

The results of water state evaluation are the characteristics as for quality and as for purity according to one of the seven categories and one of the five classes – as for quality (Table 1) and as for purity (Table 2).

TABLE 1. The categories and classes as for quality (Ukraine methodics of ecological evaluation of the surface water quality for the corresponding categories)

<table>
<tr><th colspan="2">Categories</th><th colspan="2">Classes</th></tr>
<tr><td>1</td><td>Excellent</td><td>I</td><td>Excellent</td></tr>
<tr><td>2</td><td>Very good</td><td rowspan="2">II</td><td rowspan="2">Good</td></tr>
<tr><td>3</td><td>Good</td></tr>
<tr><td>4</td><td>Moderate</td><td rowspan="2">III</td><td rowspan="2">Satisfactory</td></tr>
<tr><td>5</td><td>Satisfactory</td></tr>
<tr><td>6</td><td>Bad</td><td>IV</td><td>Bad</td></tr>
<tr><td>7</td><td>Very bad</td><td>V</td><td>Very bad</td></tr>
</table>

TABLE 2. The categories and classes as for purity (Ukraine methodics of ecological evaluation of the surface water quality for the corresponding categories)

<table>
<tr><th colspan="2">Categories</th><th colspan="2">Classes</th></tr>
<tr><td>1</td><td>Very pure</td><td>I</td><td>Very pure</td></tr>
<tr><td>2</td><td>Pure</td><td rowspan="2">II</td><td rowspan="2">Pure</td></tr>
<tr><td>3</td><td>Quite pure</td></tr>
<tr><td>4</td><td>A little pure</td><td rowspan="2">III</td><td rowspan="2">Polluted</td></tr>
<tr><td>5</td><td>Moderately polluted</td></tr>
<tr><td>6</td><td>Muddy</td><td>IV</td><td>Muddy</td></tr>
<tr><td>7</td><td>Very muddy</td><td>V</td><td>Very muddy</td></tr>
</table>

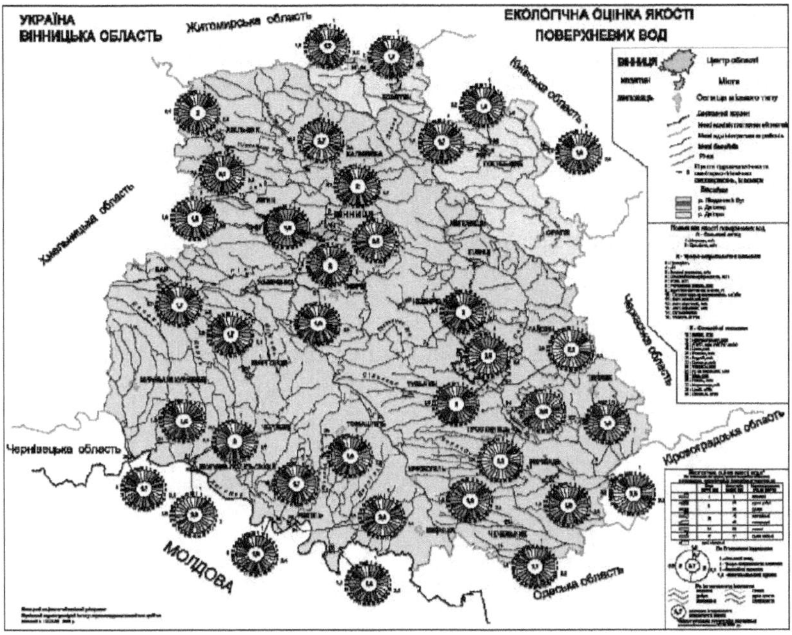

Figure 3. The automatic evaluation and forming thematic map of the region water quality according to Ukraine ministerial standards

Figure 4. The fragment of the thematic map of the region water quality according to Ukraine ministerial standards

We developed the software for the automatic evaluation and forming thematic map according to ministerial standards (Figures 3, 4).

Now we work over the creation of the alternative algorithm for the water quality evaluation on the basis of fuzzy sets theory.

Alternative algorithm consists of the following stages:

The water quality is linguistic variable.

The each i water quality category (i = 1, …, N) is the term u_i "Excellent" or "Very good," etc. with corresponding membership function $\mu(u_i)$ (Figure 5).

Many water quality factors are measured on the each observation station, its values are evaluated by one from these terms and the total characteristics of water quality is forming on all factors, for example as Figure 6.

The average membership function is forming and to conduct the defuzzification of it in according with famous formula for average water quality characteristics a (Zadeh, 1973):

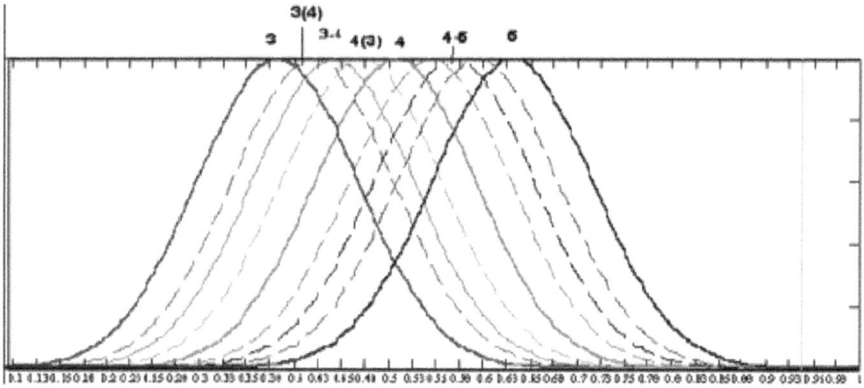

Figure 5. Membership function of water quality category term

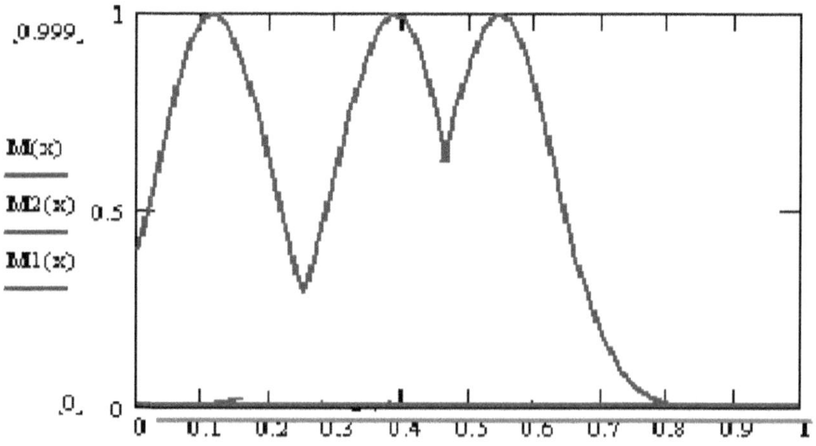

Figure 6. The example of terms of many water quality factors are measured on the one observation station

$$a = \frac{\sum_{i=1}^{N} u_i \mu(u_i)}{\sum_{i=1}^{N} \mu(u_i)},$$

for example, as for Figure 6:

$$a = \frac{0.1 \cdot 0.75 + 0.2 \cdot 0.43 + 0.3 \cdot 0.51 + 0.4 \cdot 1 + 0.5 \cdot 0.88 + 0.6 \cdot 0.64 + 0.7 \cdot 0.15}{3.76}$$

$$= 0.43.$$

The conclusion about water quality in numeric or linguistic values is made, for example, as for Figure 6: 0.43 or “Almost moderate.”

The suggested approach is more flexible in conducting the water quality evaluation, then the typical approach with the simple calculation of the average on the each of four main groups of water qualities factors.

1.1.3. *Monitoring of the nature water discharges and levels*

Besides the monitoring of the water quality, the monitoring of the nature water discharges and levels is also conducted. This is especially actual during the flood periods. This information is collected and processed daily by the employees of the Center of hydrometeorology of the Vinnytsia region.

Data entering on the data bank are conducted daily by the employees of Hydrometeocenter of the Vinnytsia Region:

- Water level
- Forecasting dangerous hydrological phenomenas
- Survey of river conditions
- Expected changes of a hydrological mode of the rivers

Hydrological bulletin is being formed daily during the flood period. Operative information is sent on the main data bank of the monitoring system.

Stages of work with our software:

Data entering and editing (MS Access, GIS Map 2000)
Data selection (MS Access)
View of the results (map, data, and text are inserted in MS Excel table for the view and print).

1.1.4. *Other abilities of the Vinnytsia region state environment monitoring*

We had created the data bank and GIS of the objects of the nature-reserve funds and places for wastes and chemical stores of the Vinnytsia region.

The build Vinnytsia region state environment monitoring system enables to form different thematic maps:

- Map of pollutant distribution on region territory
- Map with pie or bar chart of factors values
- Chart of water quality along the river
- Complex monitoring for many factors of water quality simultaneously

Important element of the environmental reporting is Web resources. We developed the Web site of the State Department of Environment and Natural Resources Management in Vinnytsia Region. This site presents the Web portal of surface water monitoring system (on Ukrainian and English) (Figure 7).

Web portal realizes the online information reports on selection observation station, basin river, or district of region for the given factor on given period. Results are displayed in the form of two-color chart: “Good quality” or “Bad quality” (see Figure 7).

We are planning to connect this Web resources to the Ukrainian map server (http://www.uamap.net). There had been signed the agreement between Vinnytsia National Technical University and the owner of this map server for using the map of Vinnytsia region.

Let us switch over to the GIS environment monitoring on the city level.

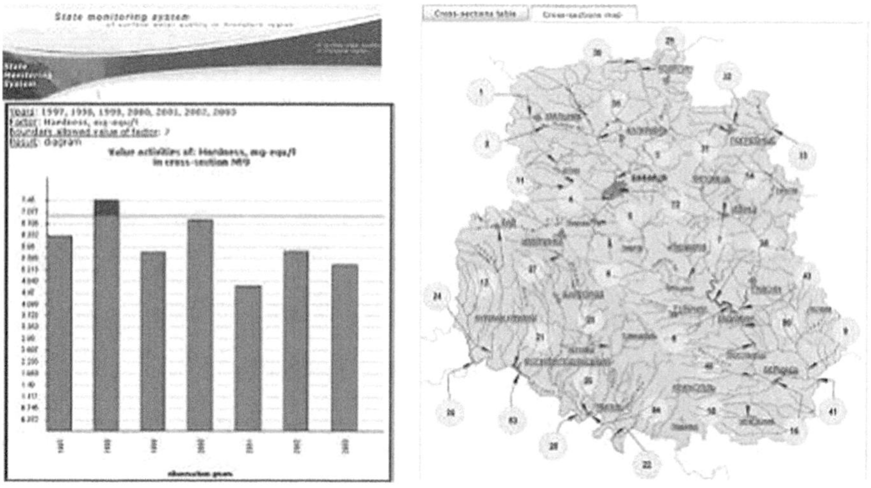

Figure 7. The Web portal of Vinnytsia region surface water monitoring system (English version) (http://www.vstu.edu.ua/vineco/netmonitoring/)

2. GIS of the State Environmental Monitoring for City Level

The map of our city is developed by the State Department of the Town-Planning and Architecture Management in Vinnytsia City (Figure 8).

We are creating the Vinnytsia city environment monitoring system in this GIS.

We created the city state monitoring of stationary sources of the air pollution (GIS soft: ArcGIS 9). We developed the map with diagram of values of NO_2, CO, and solid substances. For example, the map on "CO" is on Figure 9.

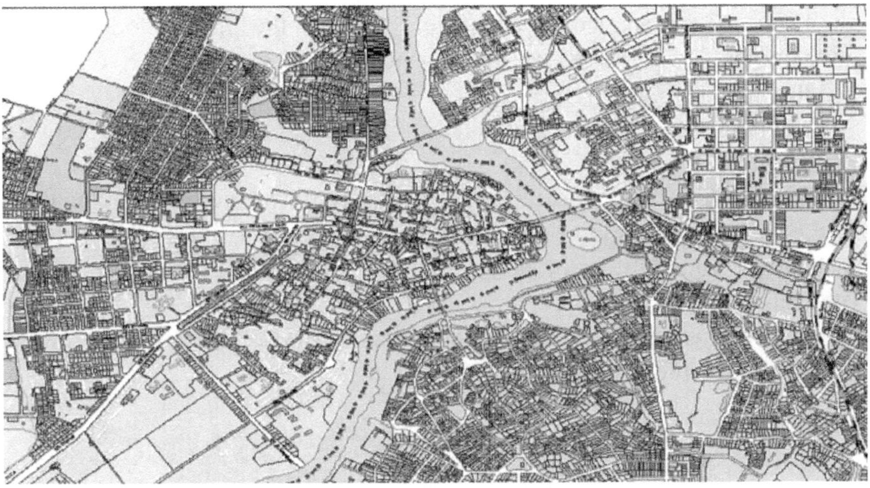

Figure 8. The map of Vinnytsia city (developed by the State Department of the Town-Planning and Architecture Management in Vinnytsia city)

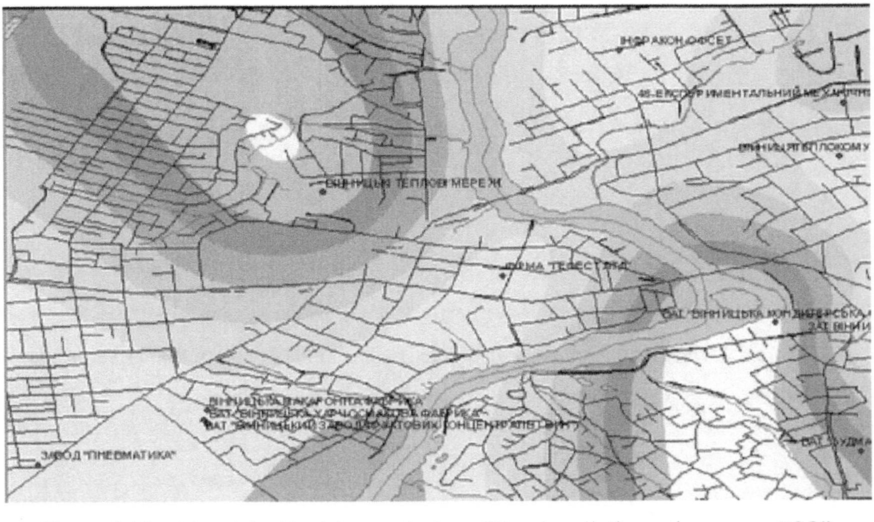

Figure 9. The Vinnytsia city state monitoring of the air pollution – the map on "CO"

3. GIS of the State Environmental Monitoring for Regional Level (many oblasts)

Let us consider the GIS environment monitoring on the level of many oblasts in particular for GIS of the Southern Booh river basin. GIS is being developed within the project "Decision Making System by Heads of the Water Management Departments in the Southern Booh River Basin with the application of the Geoinformation Technologies." The customer of this projects is the State Committee of Water Management of Ukraine.

We created the schematic map with all the streams in the basin (Figure 10).

They all are connected with each other in the graph model. All the developments for the water monitoring system in Vinnytsia region are adapted for this bigger region.

The diagram of the water quality in the river Southern Booh is shown on Figure 11.

We developed the algorithms and software for the automated calculating of the river data passport on the geoinformation map (Figure 12). It allowed to conduct calculations of the previously absent data on Southern Booh river basin.

Now, let us consider the characteristics of the created GIS environment monitoring on the country level.

Figure 10. The schematic map with all the streams in the Southern Booh river basin

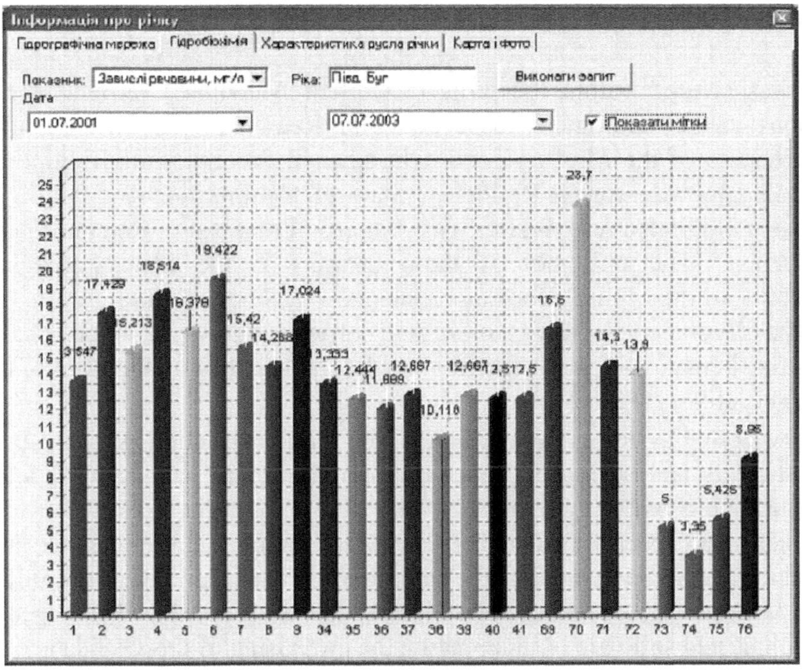

Figure 11. The diagram of the water quality in the river Southern Booh

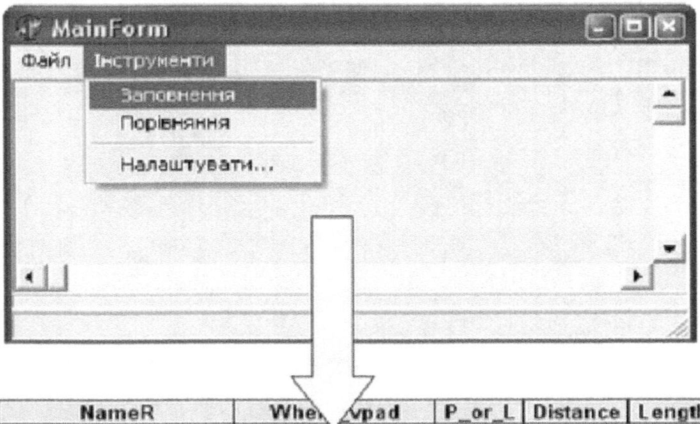

CodeR	NameR	Whe[illegible]vpad	P_or_L	Distance	Length	Zvuv
130	Р.КАЛИГУРКА	Р.ВЕЛИКА ВИСЬ	Права	90,1	13,5	1
310	Р.ЖЕРДЬ	Р.ДЕСНА	Права	57,8	21,8	1,4
54	Р.ЧИЧИКЛЕЯ	Р.ПІВДЕННИЙ БУГ	Права	642,1	128,9	1,1
364	Р.БУЖОК-41	Р.БУЖОК	Права	14,2	6,2	1,2
365	Р.БУЖОК-45	Р.БУЖОК	Права	10,3	4,9	1,1
263	Р.КАЛЬНИКА	Р.СОБ	Права	50,8	15,2	1,2
258	Р.БІЛКА	Р.СОБ	Права	65,6	12	1
239	Р.БЕРЕЖАНКА	Р.БЕРЛАДИНКА	Права	21,3	24,8	1
256	Р.КУНКА	Р.СОБ	Права	82,5	11,1	1
255	Р.КУБЛИЧ-39	Р.КУБЛИЧ	Права	20,1	12,3	1,1
290	Р.КРАСНЯНКА-1	Р.КРАСНЯНКА	Права	23,1	17	1,2

Figure 12. The automated calculating of the river data passport on the geoinformation map

4. GIS of the State Environmental Monitoring for Country Level

GIS was created within the project "Unified Automated System of the Regional State Ecological Inspection and Regional Analytical Control Departments of the Ministry for Environmental Protection of Ukraine on Obtaining the Measuring Results of State of the Environment Pollution, Emissions, Discharges, Wastes; its Saving, Processing, Retention and Analyzing." The customer of these projects is the State Ecological Inspection of the Ministry for Environmental Protection of Ukraine.

The developed geoinformation and software allows to automatically draw the thematic maps and to visualize the statistic data on the level of the country's oblasts (Figure 13).

The above systems are created by the staff, consisting of more than 100 students and scientists, under my supervision in Vinnytsia National Technical University.

Thus, had been presented the characteristics of the created and introduced GIS on the level of the city, oblast, country, and the basin of the river, which flows in seven oblasts. Also had been developed the special algorithms and software for automating the processing of this GIS data.

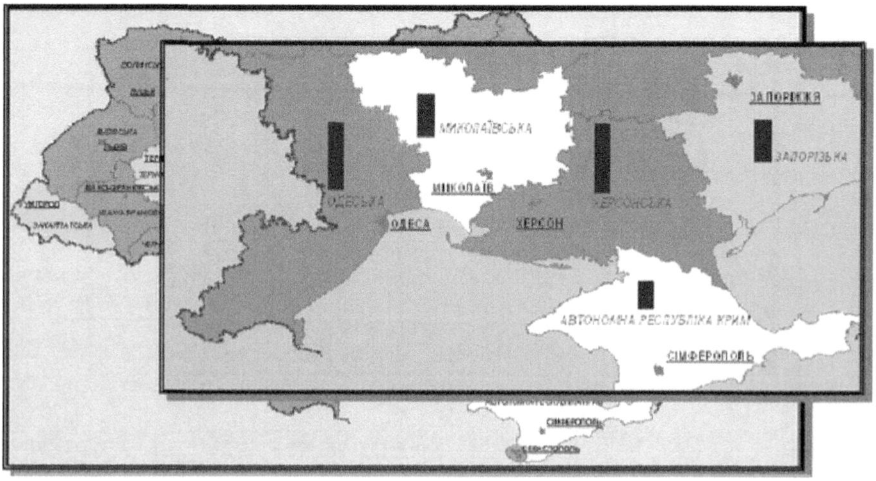

Figure 13. The thematic map with the statistic data on the level of the country's oblasts

References

V.B. Mokin, editor, *The Computer Regional Systems of Surface Water State Monitoring: Models, Algorithms, Programs*, Universum-Vinnytsia, Vinnytsia (2005).

V.B. Mokin, B.I. Mokin, Control over volume and quality of sewage water in the river waterway, *XVII IMEKO World Congress – Metrology in the 3rd Millennium. Proceedings.* TC19, 2090–2093, Dubrovnik, Croatia (2003).

L.A. Zadeh, *The Concept of a Linguistic Variable and its Application to Approximate Reasoning*, Elsevier, New York (1973).

MAPPING TYPE 2 CHANGE IN FUZZY LAND COVER

PETER FISHER*
Department of Information Science, City University, Northampton Square, London EC1V 0HB, United Kingdom

CHARLES ARNOT
Department of Geography, University of Leicester, Leicester, LE1 7RH, United Kingdom

Abstract. Discussion of the logic underlying the fuzzy change matrix is extended in this chapter to incorporate a consideration of type 2 fuzzy sets. Type 2 sets are parameterized from the differences in fuzzy set membership yielded by multiple values of the fuzziness or overlap parameter of the well-known fuzzy *c*-means classifier. Type 2 fuzzy sets can be seen as a response to the philosophical issue of higher order vagueness. It is shown that the type 2 analysis yields a variety of answers to any query about the amount of change, reflecting the higher order vagueness in the uncertainty of the change in a vague phenomenon.

Keywords: bounded difference, forest, fuzzy change analysis, fuzzy sets, fuzzy logic, land cover mapping, savanna, type 2 fuzzy sets

1. Introduction

The use of fuzzy sets in the analysis of geographical information has received considerable attention since Robinson and Strahler (1984) first suggested that this was an appropriate formal method of representation, particularly for land covers mapped from satellite imagery. Since then fuzzy memberships have been determined by a variety of methods including the Semantic Import Model relying on expert opinion, and the Similarity Relation Model using statistical clustering and pattern recognition (Robinson, 1988). One of the most commonly used fuzzy classification methods and, that used originally by Robinson and Thongs (1986), is the Fuzzy *c*-Means (FCM) classifier introduced by Bezdek (1981; Bezdek et al., 1984).

---

*To whom correspondence can be addressed: email p.fisher@city.ac.uk

A. Morris and S. Kokhan (eds.), Geographic Uncertainty in Environmental Security, 167–186.

In common with much work in fuzzy sets, however, research in the fuzzy interpretation of satellite imagery to date has focused on type 1 fuzzy sets. In this paper we extent the analysis into a type 2 fuzzy analysis. The paper starts by explaining the basis of type 2 fuzzy sets and suggests an approach by which they may be parameterized from the FCM classifier. In section 3 we revisit the fuzzy logic of change analysis and extend this change analysis to type 2 sets. The results and conclusion sections outline the advantages of a type 2 analysis and considers some ways in which the results of type 2 analysis can be presented.

2. Type 2 Fuzzy Sets and Higher Order Vagueness

Fuzzy set theory is a deliberate attempt to accommodate vagueness of class, object, process, or location (collectively phenomena) definition in set theory (Fisher, 2000). As such it builds on the writings of Russell (1923), Black (1937), and Kaplan and Schott (1951). Unlike these philosophers, Zadeh (1965) wrote for a technology audience which was receptive to the ideas of phenomena only partially belonging to sets and wished to see such solutions implemented in information systems. As a result the fuzzy sets and fuzzy logic articulated by Zadeh (1965) started an avalanche of research in the area. The vast majority of that research has addressed only the direct form of vagueness – first order vagueness. A measurement of some type is made, and a membership function derived for the degree to which it is possible to infer the phenomenon's membership of a particular set from that measurement; for example, the height of a person is related to the membership of the set of tall people, or the proportion of oak trees in a woodland to the membership of the set of oak woodlands.

The problem with this is that, given a membership function for any particular measurement there is one specific membership value. In other words there is no doubt about the fuzziness or degree of membership, once the fuzziness (membership function) is defined. This is a paradoxical situation which is most clearly illustrated in fuzzy set theory by the concept of α-cuts. For any threshold fuzzy membership (α), a crisp, Boolean set is created from the fuzzy set by allowing any phenomenon with a membership greater than α to be part of the α-cut (see for example, Arnot et al., 2004). In short the fuzzy set is crispened to the α-cut which can then be treated as a Boolean set. But if a statement about the set (the α-cut) is Boolean, then the set is Boolean! Perhaps there should be a residual doubt or vagueness as to the membership of an object in the α-cut set. In the philosophical writing on vagueness, this residual doubt is the concept of higher order vagueness. In summary, if it is possible to make a precise statement about a vague phenomenon then that phenomenon is not vague. To consider the phenomenon truly vague, it is necessary that any statement that can be made about it should itself be vague (Sorenson, 1985). In fuzzy set terms this means that for any α-cut through a fuzzy set, a fuzzy set should result.

In the vagueness literature, there has been a long-running debate on whether higher order vagueness is a necessary property of vague phenomena. Although there is no consensus on the issue, it is widely accepted that higher order vagueness is possible (Varzi, 2003). Therefore it should be accommodated in any theory which is supposed to address vagueness.

For fuzzy sets this issue was first recognized by Zadeh in a paper published in 1975 (ten years before Sorenson's paper) and termed type n fuzzy sets. The mathematics and computational overhead of working with type n fuzzy sets were considered too costly at the time, and very few studies followed up the topic. Eventually computing power caught up with the problem and with refinement of the method there has been a resurgence of interest in type 2 fuzzy sets in particular (Mendel, 2001; Mendel and John, 2002). Higher order fuzzy sets still tend to be considered either beyond the competence of computing and conceptualization or to be unnecessary.

In geographical information processing, there have been few discussions of type 2 fuzzy sets. Verstraete et al. (2005) mention type 2 sets as part of their data model, but they do no more than that, and Kulik (2003) discusses the issue of second order uncertainty in terms of supervaluation theory. Fisher et al. (in press) present a full justification of type 2 spatial fuzzy sets and a preliminary analysis showing some advantages of the approach. They discuss the recognition of the vague object a mountain peak, and they show that it is possible to construct type 2 sets for this object. They examine the visibility of the peaks concerned, and demonstrate that for any location it is possible to derive upper and lower bounds on the possibility of the peak being visible; a novel and potentially useful result in landscape planning. The type 2 fuzzy sets are found by examining the alternative values for two key parameters of an algorithm for peak detection; varying values of one parameter yields the type one fuzzy set (Fisher et al., 2004) and varying the second the type 2 fuzzy set (Fisher et al., in press). A similar approach is used here.

2.1. TYPE 2 FUZZY SETS AND FUZZY *C*-MEANS

In fuzzy c-means classification one key parameter in the algorithm is the fuzziness, m (Bezdek, 1981; Bezdek et al., 1984), fuzzy exponent, β (Deer and Eklund, 2002), and fuzzy overlap, q (de Gruijter and McBratney, 1988), which are all related, and referred to as m henceforth. This value ranges from 1 to infinity (theoretically). In the special case when m = 1, the classification is crisp, and the result of a fuzzy *c*-means classification is the same as the normal *k*-means clustering. As *m* increases, so the values of memberships converge until when *m* is large the memberships in all classes for all cases are equal.

Selection of *m* has exercised the minds of users of the FCM classification procedure. Bezdek et al. (1984) state that "for most data, 1.5 $\leq m \leq$ 3.0 gives good results". On the other hand, McBratney and Moore (1985) found 2 to be "optimal", while Chloe and Jordan (1992) suggested 12 was optimal. Deer and Eklund (2003, 201) used 1.6, and Fisher et al. (2006) state that "the usual value of 2 for fuzziness was found acceptable". Statements of this type abound in the literature. Although many researchers use variants of the Partition Coefficient (Bezdek et al., 1984) to determine the optimal number of clusters extracted from a data-set, an approach to assessing the optimality of the classification resulting from various values of *m* have not been suggested until recently. Okeke and Karnieli (2006) have investigated values of *m* from 1.1 to 2.5 in steps of 0.1 and suggested a new measure of optimality.

In the research reported here, however, instead of searching for an optimal value of *m* we treat a number of values as containing equally useful information and use that information as a basis for parameterizing the type 2 fuzzy set. Following approximately in the steps of Okeke and Karnieli (2006) we examine classification results for $m = 1.3$ to 2.5 in steps of 0.1.

Given that a number of different valuations of *m* are used, they can be used to derive parameters of a type 2 fuzzy set just as Fisher et al. (in press) parameterize type 2 fuzziness of peaks. By summarizing the fuzzy memberships of the group of type 1 sets, it is possible to derive parameters of a type 2 set, as in Equations 1 to 3.

$$\mu_{2\max}(X) = \max_{j=1.3}^{2.5} \left(\mu_j(X)\right) \qquad (1)$$

$$\mu_{2\min}(X) = \min_{j=1.3}^{2.5} \left(\mu_j(X)\right) \qquad (2)$$

$$\mu_{2\text{mean}}(X) = \sum_{j=1.3}^{2.5} \frac{\mu_j(X)}{n} \qquad (3)$$

Where ***j*** takes values of *m* from 1.3 to 2.5 in steps of 0.1 giving, in this instance $n = 13$. From $\mu_{2\min}(X)$ and $\mu_{2\max}(X)$ (Equations 1 and 2) it is possible to define an upper and lower bound for the fuzzy membership of a class. The minimum is the smallest membership value at a location, and effectively maps out those areas which show the highest affinity with the concept of the class. The maximum is the largest possible fuzzy membership, and shows the degree of membership in the most generous interpretation of the class. $\mu_{2mean}(X)$ (Equation 3) gives a third approximation which might be termed the typical membership of the class. The three values, $\mu_{2min}(X)$, $\mu_{2mean}(X)$, $\mu_{2max}(X)$ can be seen as type 2 α-cuts through the fuzzy membership function of class X. This can be interpreted as a three-value membership function of the form $\mu_2(X) = (\mu_{2min}(X), 1, \mu_{2mean}(X), 0.5, \mu_{2max}(X), 0)$, where 1, 0.5, and 0 are all taken to be approximate valuations.

3. Fuzzy Change in Land Cover

The full expression of the fuzzy change matrix has been articulated by Fisher et al. (2006). The diagonal cells of the change matrix indicating no change has occurred are given by the usual fuzzy intersection (minimum) between the fuzzy memberships of that cover type at time t_1 and time t_2.

$$\mu(X_i)_{[static]} = \min \left(\mu(X_i t_1), \mu(X_i t_2)\right) \tag{4}$$

For off-diagonal elements it is necessary to first define the gain and loss of each cover type using the alternative fuzzy intersection operator, the bounded difference (see Fisher et al., 2006 for full justification; Klir and Yuan, 1995). That is to say the Gain of X_i is thoses areas which are not X_i at t_1 AND X_i at t_2

$$\mu(X_{i[gain]}) = \max \left(0,\ \mu(X_i \neg t_1) + \mu(X_i t_2) - 1\right) \tag{5}$$

On the other hand, the loss in X_i is given as those area which are X_i at t_1 and not C_i at t_2.

$$\mu(X_{i[loss]}) = \max \left(0,\ \mu(X_i t_1) + \mu(X_i \neg t_2) - 1\right) \tag{6}$$

The off-diagonal values for the change matrix are then determined as the intersection of the loss of cover type i and the gain of cover type j, where i ≠ j

$$\mu(X_i X_j)_{[change]} = \min \left(\mu(X_i)_{[gain]}), \mu(X_j)_{[loss]}\right) \tag{7}$$

In any analysis of type 2 fuzzy sets there is a potential to generate much data (one of the features that put off early work in this area). Using distributed spatial models, as here, that data is multiplied by the number of locations modelled (the number of pixels). Therefore it is necessary to summarize the results of analysis. The principal measure used here, the fuzzy area, is determined from the fuzzy cardinality (Verstraete, this volume), which for a raster image of the fuzzy set X having dimensions r rows by c columns is given as:

$$Fuzzy\_Area_{\mu(X)} = \sum_{l=1}^{c} \sum_{k=1}^{r} \left(\mu(X)_{l,k} \right) \tag{8}$$

3.1. TYPE 2 FUZZY CHANGE

The simplest form of type 2 fuzzy analysis is to examine the results of the analysis in each instance of the type 1 fuzzy set and to summarize those instances using Equations 1, 2, and 3 to give the triangular type 2 fuzzy set. Therefore in this study we examine the properties of the type 2 fuzzy sets of separate land covers at two different dates, by examining all instances of those land covers from $m = 1.3$ to $m = 2.5$, and for completeness we present aspects of the type 2 sets of each cover type at each date. We then execute the change analysis for each value of m, and present type 2 sets of the possibility of change.

4. Study Area

Following Arnot et al. (2004) and Fisher et al. (2006), we examine changes in a small portion of the environs of the savanna–forest boundary in central Bolivia. The area is typified by two principal land cover types: the savanna and the forest. Within the area water is a major controlling influence on both covers. There is standing water in lakes and rivers, but there are also wet and flooded forest areas around watercourses, and the savanna becomes inundated during the wet season and can itself be separated into wet and dry savanna according to both species present and the wetness of the soil and health of the plants. The forest is quite well defined conceptually, as is the water and savanna. Each can be expected to blend into each other spatially yielding vague boundaries. The separation of savanna types is much harder. In this paper we again study a small portion of the area studied by Arnot et al. (2004) and a different portion than examined by Fisher et al. (2006).

Two Landsat TM images were analysed for this study, one from the dry season in 1985 and one from the wet season in 1986. They were acquired for a related project to generate traditional Boolean maps of the vegetation of the area (Millington, 1996). Fuzzy classification was executed with the Parbat software (Lucieer, 2004), which includes implementations of both the unsupervised or supervised versions of the FCM classification. From experimentation with both dates, it was found that four classes were identifiable in both years which, from knowledge of the landscape, could be associated with the four named classes above, forest, water, and wet and dry savanna.

This study is principally concerned with conclusions which relate to expanding the representational variety of geographical information; can type 2 sets be identified from fuzzy classification? Does type 2 fuzzy analysis add anything new to the results over a type 1 analysis? Any conclusions related to change in the landscape should be treated with caution at this stage, however, and form the focus for further investigation.

5. Results

5.1. TYPE 2 SETS OF LAND COVERS

The sequence of fuzzy membership images (maps) of the study area in Figure 1 shows an increasing fuzzification of the forest as the fuzzy exponent (fuzziness) is increased. The maximum membership of forest in the area is constant (there are white areas in all images). The extent in the maps of areas with some affinity (degree of membership greater than zero) to forest, however, increases (more areas are shown in a shade of grey), and the maps show an increasing structure in the pattern of the fuzzy membership (reflected in the variation of greys). Similar detailed patterns are revealed in the corresponding set of maps for other cover types, which are masked by the degree of crispness in maps with smaller fuzziness.

Table 1 lists the percentages of the area occupied by all cover types in the two years, tabulated against the fuzziness. Any one fuzzy analysis (either map or row in the table) might be chosen as representative, depending on the judgement of the investigator, but a type 2 analysis takes a different view.

In Table 2 the percentage of the study area occupied by the triangular type 2 fuzzy sets based on Equations 1, 2, and 3 for each land cover are presented. Note that none of the values in this table are included in Table 1. At any pixel the values are summarized across valuations of m, so for any one pixel, the minimum memberships, for example, of type 1 sets are taken, and will not necessarily be from the classification with the same value of m at every pixel.

Figure 2 illustrates transects through all 13 of the type 1 set memberships of water (Figure 2A) and wet savanna (Figure 2C) and the equivalent type 2 fuzzy sets for (Figures 2B and 2D respectively). As in the greyscale images (Figure 1) a subtle structure can be seen from the variety of type 1 membership values at any location and their variation along the transect. This variation is reflected in the type 2 memberships, although because they are summaries of the results from the 13 values of m (type 1 sets), they do not show the subtlety of the original type 1 sets. Specifically, a zone of wet

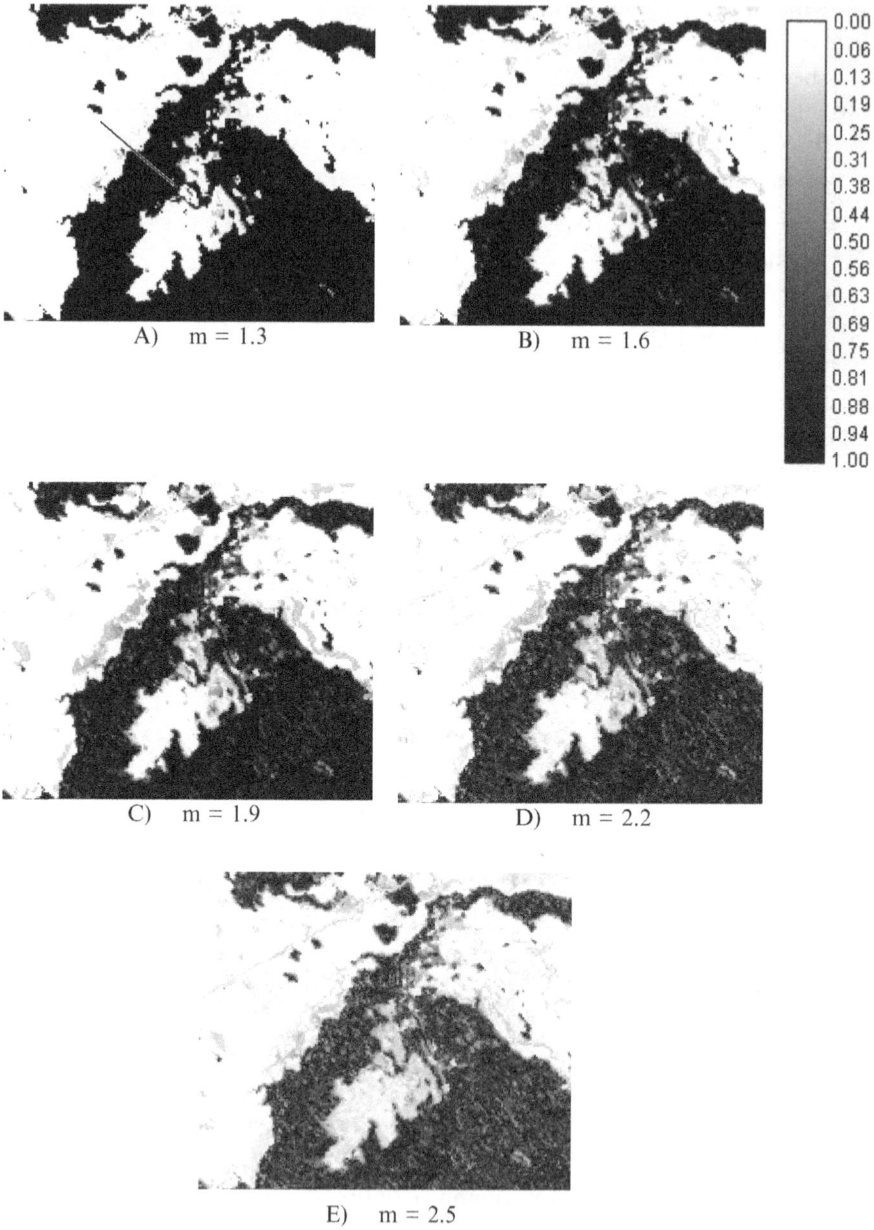

Figure 1. Results for five different valuations of the fuzziness exponent in the FCM classifier for the forest land cover

savanna within the transect which is almost a perfect Boolean class in the 1.3 fuzziness (m) type 1 fuzzy memberships, degrades through the valuation of fuzziness so that parts of it are below 0.2 membership when $m = 2.5$. The

TABLE 1. The percentages of the study area which are occupied by each of four land covers in the two years studied and as a result of 13 different valuations of the fuzzy exponent

		1985		
Fuzziness	Water	Wet savanna	Dry savanna	Forest
1.3	0.87	14.14	39.33	45.66
1.4	0.88	15.00	38.54	45.58
1.5	0.92	16.03	37.65	45.41
1.6	1.01	17.13	36.74	45.12
1.7	1.19	18.26	35.85	44.70
1.8	1.44	19.35	35.04	44.18
1.9	1.77	20.36	34.30	43.57
2	2.15	21.30	33.64	42.90
2.1	2.59	22.13	33.08	42.19
2.2	6.68	23.12	30.18	40.03
2.3	9.24	23.37	28.90	38.49
2.4	10.87	23.58	28.19	37.36
2.5	12.29	23.72	27.64	36.36
		1986		
Fuzziness	Water	Wet savanna	Dry savanna	Forest
1.3	1.93	23.78	24.67	49.62
1.4	2.00	23.77	24.81	49.42
1.5	2.18	24.17	24.55	49.11
1.6	2.46	24.63	24.22	48.68
1.7	3.32	25.36	23.77	47.55
1.8	3.32	25.36	23.77	47.55
1.9	3.85	25.58	23.69	46.87
2	4.43	25.78	23.65	46.14
2.1	5.05	25.94	23.66	45.36
2.2	5.68	26.06	23.70	44.56
2.3	6.33	26.16	23.76	43.75
2.4	6.97	26.25	23.83	42.94
2.5	7.61	26.32	23.36	42.15

fuzzy membership released (as it were) when m is increased can be seen to be partly taken up by the associated increase in the membership of water with m in this zone.

Transects through the fuzzy memberships of the forest class (not illustrated) also show an increased membership with m. Thus a classification with large m shows the zone to be a mixed area of water, wet savanna and,

TABLE 2. The percentages of the study area which are occupied by the maximum mean and minimum (triangular type 2 set) of the four land covers in the two years studied

		1985		
Fuzziness	Water	Wet savanna	Dry savanna	Forest
Maximum (generous)	12.72	30.66	44.85	50.67
Mean (typical)	3.99	19.81	33.77	42.43
Minimum (core)	0.47	7.51	22.41	31.93
		1986		
Fuzziness	Water	Wet savanna	Dry savanna	Forest
Maximum (generous)	8.05	34.17	32.31	54.09
Mean (typical)	4.24	25.32	24.00	46.44
Minimum (core)	1.51	16.09	16.33	37.70

to a lesser degree, forest, although when m is small it is identified as almost quintessential wet savanna. With increase in m it is a property of the FCM classifier that memberships will converge towards $1/c$, but in the range 1.3 to 2.5, it can be seen in the transects that very few locations achieve this small value. Rather some classes can rise from a very small membership to a quite large membership when other classes fall.

These variations of membership are reflected in the minimum–maximum envelope of the type 2 fuzzy set, and detail is added by the mean value of the 13 type 1 sets. In short, as suggested in discussion of Figure 1, a subtle structuring of the landscape (variation in grey scale values in Figure 1 and along the transect in Figure 2) is revealed in the classification results. The suggested type 2 treatment would appear to reflect underlying information about the landscape which is not revealed by any single type 1 valuation. The range of values of membership reportable for a location (as revealed in the transects or maps) or summarized across the landscape (Tables 1 and 2) show that the type 2 representation would appear to express a variety of possible results as is desirable.

5.2. TYPE 2 SETS OF CHANGED LAND COVERS

If it is possible to suggest that a type 2 set can be represented for each land cover as in the preceding section, then it is also possible to determine the change in cover. For each valuation of m, therefore, gain and loss of each cover types was determined and the fuzzy change matrix populated, according to Equations 4 to 7.

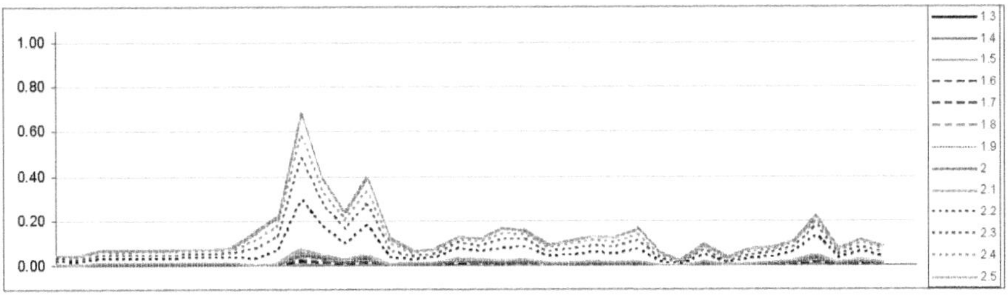

(A) Type 1 fuzzy sets of water

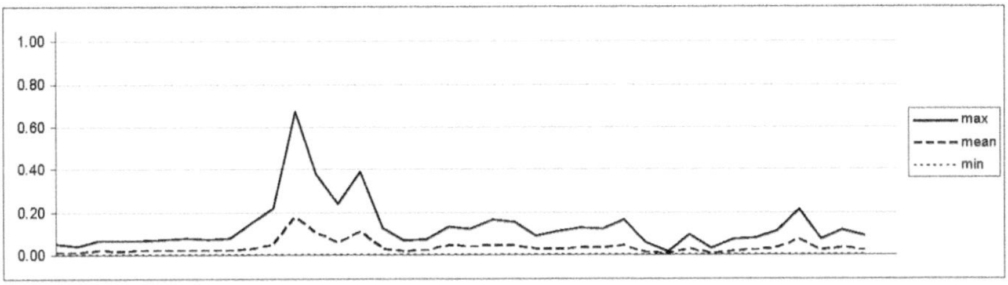

(B) Type 2 fuzzy sets of water

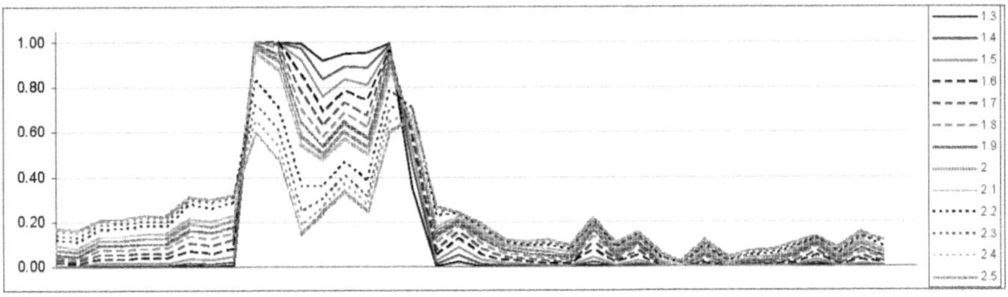

(C) Type 1 fuzzy sets of wet savanna

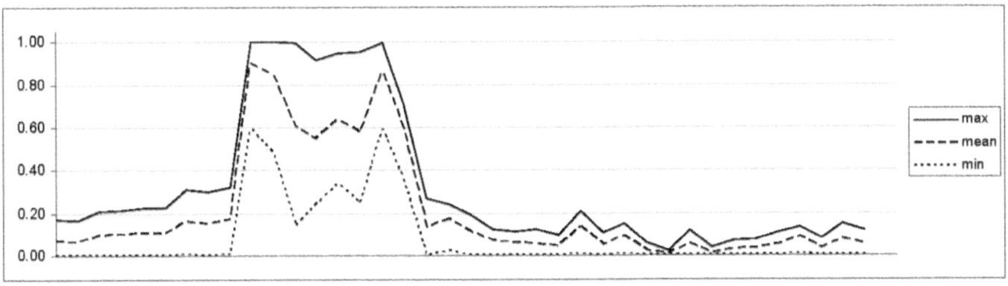

(D) Type 2 fuzzy sets of wet savanna

Figure 2. Variation of type 1 and type 2 fuzzy sets along a 1,500 m transect of the study area (position of transect shown in Figure 1A)

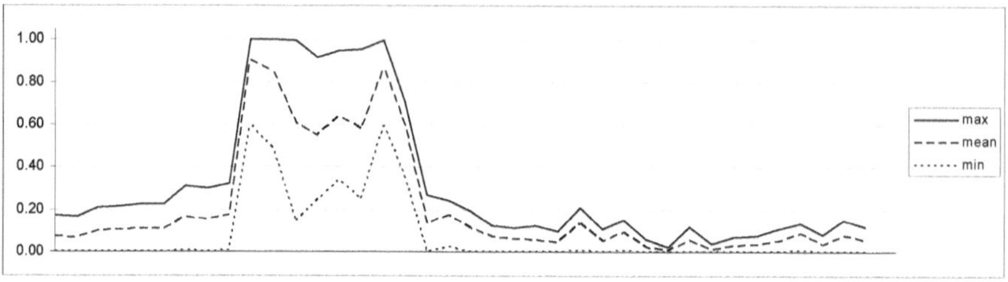

(A) Type 2 fuzzy sets of wet savanna in 1985

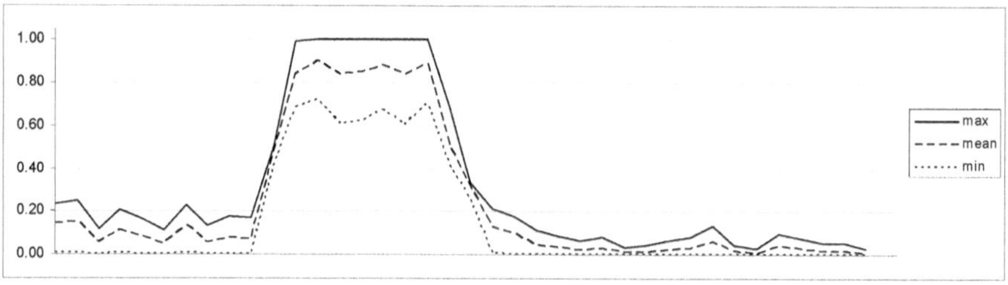

(B) Type 2 fuzzy sets of wet savanna in 1986

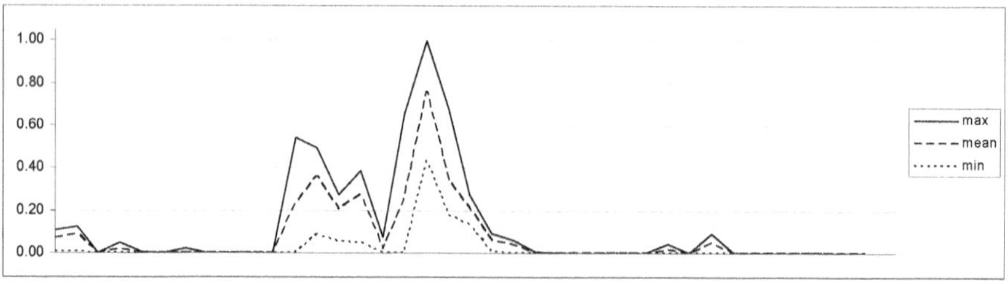

(C) Gain of type 2 fuzzy sets of wet savanna 1985–1986

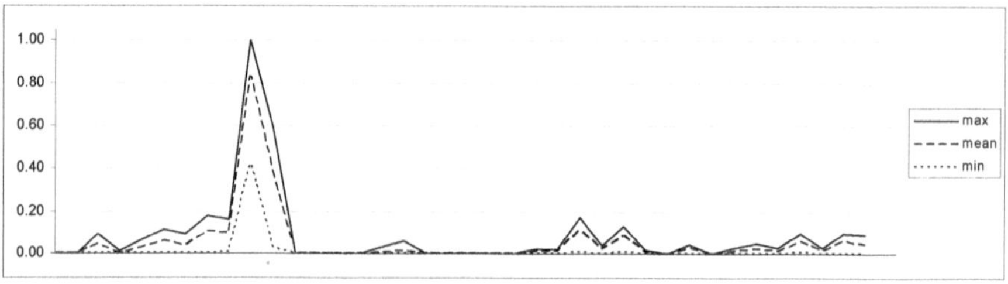

(D) Loss of type 2 fuzzy sets of wet savanna 1985–1986

Figure 3. Variation of type 2 fuzzy sets and fuzzy change sets for wet savanna along a 1,500 m transect of the study area (position of transect shown in Figure 1A)

This, of course, results in 13 different measures of gain and loss of each cover type and 13 different change matrices. In Figure 3 transects (along the same as shown in Figure 2) show not only the type 2 fuzzy sets of wet savanna in 1985 and 1986, but the gain and the loss of each. The extremely

TABLE 3. The percentages of change from and to each cover type (the fuzzy change matrix) in the study area for five valuations of *m*

m = 1.3			1986	
1985	Water	Wet savanna	Dry savanna	Forest
Water	0.12	0.40	0.22	0.14
Wet savanna	0.73	7.21	3.74	2.47
Dry savanna	0.91	13.76	19.08	5.59
Forest	0.18	2.43	1.64	41.42
m = 1.6			1986	
1985	Water	Wet savanna	Dry savanna	Forest
Water	0.26	0.34	0.25	0.21
Wet savanna	0.82	9.54	3.76	3.05
Dry savanna	1.07	12.03	18.40	5.28
Forest	0.37	2.79	1.86	40.17
m = 1.9			1986	
1985	Water	Wet savanna	Dry savanna	Forest
Water	0.89	0.32	0.26	0.37
Wet savanna	1.01	12.60	3.40	3.49
Dry savanna	1.41	9.87	18.10	5.07
Forest	0.71	2.92	2.03	38.05
m = 2.2			1986	
1985	Water	Wet savanna	Dry savanna	Forest
Water	3.04	1.24	0.91	1.72
Wet savanna	0.96	15.53	3.29	3.47
Dry savanna	1.25	6.85	17.71	4.51
Forest	0.57	2.68	2.01	35.00
m = 2.5			1986	
1985	Water	Wet savanna	Dry savanna	Forest
Water	5.62	2.21	1.72	3.27
Wet savanna	0.75	17.33	2.75	3.04
Dry savanna	0.98	4.88	18.00	3.96
Forest	0.37	2.38	1.90	32.11

localized peak in the loss of forest does suggest the possibility of a geometric misregistration between the different date images. It is matched by a corresponding peak of gain on the other side of the clearing, although the fact that this is a wider peak does suggest there may be more substance to the gain detected.

In Table 3 the five fuzzy change matrices for the same values of m as are illustrated in Figure 1. First it should be noted that these are percentages and all matrices sum to 100% although only approximately due to some rounding occurring. The main differences in the amount of change are associated with locations where no change of cover has occurred. The fuzzy area of no change in water displays the most dramatic range; the difference between the least area and the most area is 0.12–5.62% of the study area, increasing with m. The largest area represented in all change matrices for valuations of m is that of no change in forest; it ranges from 32.11% to 41.42% of the study area, with decreasing values of m.

TABLE 4. The percentages of the maximum, mean and minimum change from and to each cover type (the type 2 fuzzy change matrix) in the study area

Maximum (generous)		1986		
1985	Water	Wet savanna	Dry savanna	Forest
Water	5.66	2.48	1.86	3.38
Wet savanna	1.61	21.11	5.61	5.20
Dry savanna	2.02	15.39	26.52	7.78
Forest	1.03	4.34	3.18	45.87
Mean (typical)		1986		
1985	Water	Wet savanna	Dry savanna	Forest
Water	1.77	0.79	0.59	0.99
Wet savanna	0.89	12.47	3.39	3.16
Dry savanna	1.19	9.54	18.25	4.91
Forest	0.50	2.70	1.92	37.47
Minimum (Core)		1986		
1985	Water	Wet savanna	Dry savanna	Forest
Water	0.07	0.16	0.12	0.07
Wet savanna	0.25	3.79	1.27	0.88
Dry savanna	0.44	4.37	10.84	2.19
Forest	0.10	1.09	0.76	27.86

The largest amount of actual change reported is of dry savanna to wet savanna (13.76% of the area) as should be expected in a dry season to wet season change. This amount of change is in the smallest valuation of m, and, as for the stable area of forest, the spatial extent of this category of change decreases as m increases.

Table 4 lists the parameters of the type 2 sets of land cover change. Interestingly, the total of the percentages reported for the maximum possible change is 153%, the total for mean change 100% and the total for the minimum change 54%. This is inevitable because there is no constraint on any parameter that the fuzzy memberships within a pixel sum to 1. Rather the memberships are evaluated only with respect to the values. The variation in total change indicates that, while we can be definite about the change condition (static or changed) of 54% of the area (minimum of the type 2 set), we can consider that while up to 153% may have changed.

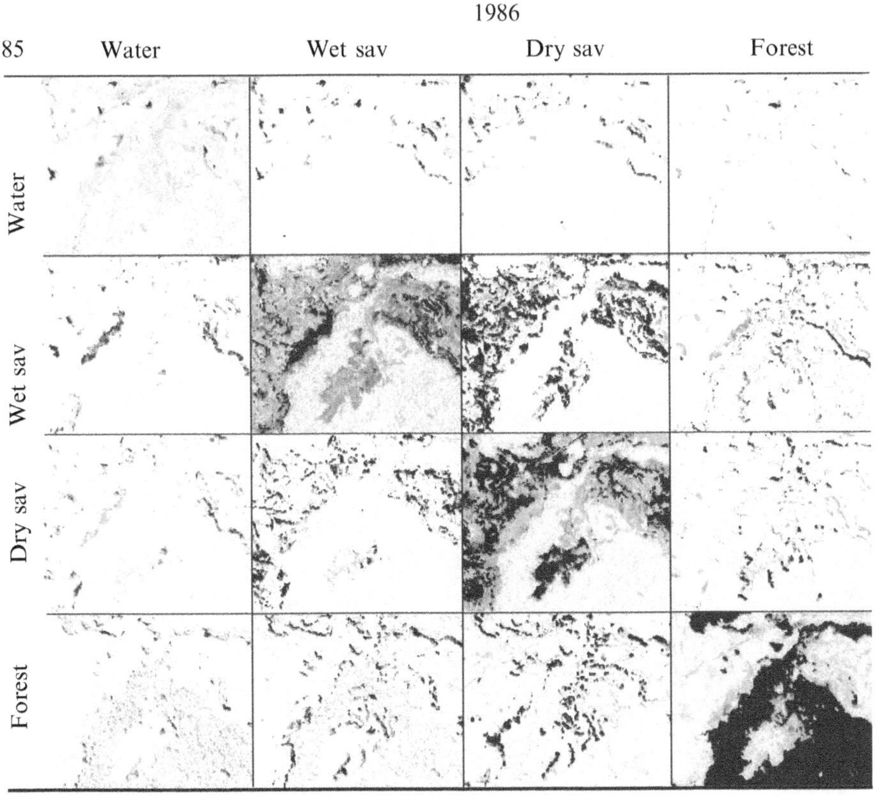

Figure 4A. Maps of the degree of maximum (generous) memberships of type 2 set of land cover change from and to each cover type. For grey-scale legend see Figure 1

Figures 4A and 4B show maps of each cell of the change matrix for the maximum and minimum values of the type 2 fuzzy change sets (the interval type 2 fuzzy set). The maps for maximum membership (Figure 4A) naturally seem to show the most definite expression, but the maps in Figure 4B which show the minimum memberships show the pixels most similar to forest. The impression given by the set of maps, and confirmed by the numerical values in Table 4, is that the principal changes between land covers in the area is between forest and savannas both in terms of presence and of change, rather than between water and anything else. Although there may have been areas of change where wet savanna became water, for example, (Figure 4A), none of the cells can be described as having a high membership in the core of this change class (Figure 4B). The same is true for all maps in the row and column associated with water in Figure 4B. Indeed the dominant definite patterns are that dry savanna and forest areas have not changed. The core extent or minimum membership (Figure 4B) of any other conditions is small, and it is hard to find good examples of their trajectory of change.

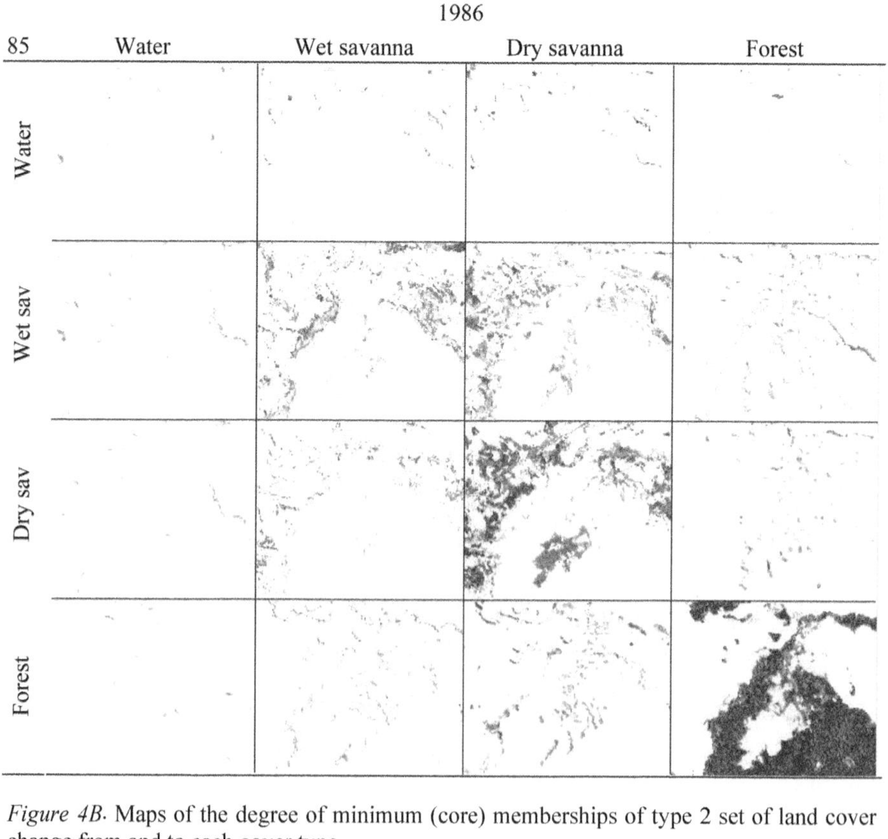

Figure 4B. Maps of the degree of minimum (core) memberships of type 2 set of land cover change from and to each cover type

Another way of interpreting the maps is to say that the areas shown in black in the minimum map (Figure 4B) are those whose status is relatively well known showing closest affinity with the classes identified, while those in the maximum map (Figure 4A) are those which have only a little affinity with the labelled classes. In other words, the maximum map warns of areas where changes in land cover type may be occurring, to only a small degree, while the minimum map shows where they are definitely occurring.

5.3. REPORTING TYPE 2 CHANGE RESULTS

The results of type 2 analysis are intended to produce a range of values for any interpretation. If we have a concept of typical forest (our core concept), but we are not sure of the extent in attribute space of forest (vagueness). It follows that the forest is well modelled as a spatial fuzzy set, whose boundary has a vague spatial extent. This being the case we can as much say that a specific area is forest as we can that one location is forest to some specific degree. This rehearses the justification of a type 2 analysis of land cover extents and change.

This argument presents a quandary for reporting the outcome of analysis of type 2 sets. It can be worrying enough for land managers familiar with simplistic Boolean mappings of the landscape to be presented with fuzzy interpretations but to be presented with type 2 fuzzy interpretations may be even more concerning.

One way to report these results, however, is to use natural language terms such as those suggested in section 2.1. to convey the meaning of the parameters of the type 2 sets. Thus we might report that in 1985 the forest may have extended to between 42.43% in a typical interpretation and 50.67% in a generous interpretation of the study area, but the core area of forest was only 31.93% of the area., and in 1986 the area may have been between a typical 46.44% and generous 54.09% but the core forest occupied 37.70%.

The gain in areas that are core forest, however, was only about 4.57% of the area at the same time as the core forest lost 2.08% of its area. In a typical interpretation of forest, the gain could have been as much as 8.96% or even 13.93% in a generous interpretation while the loss could have been 4.95% for the typical or as much as 7.50% for the generous.

To be more specific, within the landscape, between 1.09% and 4.34% of the forest became more similar to wet savanna although most typical of those classes in 1985 and 1986 was 2.7% of the area.

Values such as these could be gleaned from α-cuts of any single type 1 fuzzy set, but which type 1 set should be chosen and which α-cuts? Using the type 2 analysis presented removes any concern about which to use. The remaining concern is exactly which range of values of m should be used in such a type 2 analysis. With only 13 instances (valuations) included in these results it cannot be doubted that changes to estimates of membership and fuzzy area would occur if another spread of values of m were used.

6. Conclusion

In this paper we have shown that using the approach of multiple valuations of classification parameters it is possible to generate multiple instances of the classification scheme. This much is a necessary property of the classifier. Unlike previous researchers, however, we have not asked which valuation yields the *correct* or *best* classification. Rather we have explored the variation in fuzzy memberships derived as an expression of type 2 fuzzy membership.

In the resulting analysis, we have then mapped areas which typify concepts (as fuzzy sets) as well as areas with more uncertainty over their affinity with the cover types named. We have gone on to show how this can be used in a change detection analysis extending earlier work on type 1 change analysis to type 2 change analysis. The results allow the identification of areas which are good representatives of the named cover (but still to a degree identified) and those that are to a degree less good representatives. Finally suggestions are made for using natural language qualifiers in presenting interpretations of these results.

We believe that type 2 fuzzy sets give more power to the analysis and interpretation of environmental phenomena because they allow for the detection of information-rich changes, possibly giving early warning of undesirable trends in succession or land cover change.

Acknowledgement

We would like to thank Richard Wadsworth and Jane Wellens for their contributions to preliminary stages of this work. The results and conclusions are entirely the responsibility of the authors.

References

Arnot, C., Fisher, P.F., Wadsworth, R., and Wellens, J., 2004, Landscape Metrics with Ecotones: pattern under uncertainty, *Landscape Ecology* **19**: 181–195.

Bezdek, J.C., 1981, *Pattern Recognition with Fuzzy Objective Function Algorithms*, Plenum Press, New York.

Bezdek, J.C., Ehrlich, R., and Full, W., 1984, FCM: The fuzzy *c*-means clustering algorithm, *Computers & Geosciences* **10**: 191–203.

Black, M., 1937, Vagueness: an exercise in logical analysis. *Philosophy of Science* **4**: 427–455.

Choe, H., and Jordan, J.B., 1992, On the optimal choice of parameters in a fuzzy *c*-means algorithm. *Proceedings of the IEEE International Conference on Fuzzy Systems*, IEEE Service Center, Piscataway, NJ, pp. 349–354.

Deer, P.J., and Eklund, P., 2003, A study of parameter values for a Mahalanobis Distance fuzzy classifier, *Fuzzy Sets and Systems* **137**: 191–213.

Fisher, P.F., 2000. Sorites Paradox and Vague Geographies. *Fuzzy Sets and Systems* **113**: 7–18.

Fisher, P.F., Wood, J., and Cheng, T., 2004, Where is Helvellyn? Multiscale morphometry and the mountains of the English Lake District. *Transactions of the Institute of British Geographers* **29**: 106–128.

Fisher, P.F., Arnot, C., Wadsworth, R., and Wellens, J., 2006, Detecting change in vague interpretations of landscapes, *Ecological Informatics* **1**: 163–178.

Fisher, P.F., Cheng, T., and Wood, J. (in press). Higher order vagueness in geographical information: Empirical geographical population of Type n fuzzy sets. *GeoInformatica*.

Gruijter, J.J. de, and McBratney, A.B., 1988, A modified fuzzy k-means method for predictive classification, in: H.H. Bock (ed.), *Classification and Related Methods of Data Analysis*, Elsevier, Amsterdam, pp. 97–104.

Kaplan, A., and Schott, H.F., 1951, A calculus for empirical classes, *Methodos* **3**: 165–188.

Kulik, L., 2003, Spatial vagueness and second order vagueness. *Spatial Cognition and Computation*, 3, 157–183.

Lucieer, A., 2004. Parbat: version 0.32. www.parbat.net.

McBratney, A.B., and Moore, A.W., 1985, Application of fuzzy sets to climatic classification, *Agricultural and Forest Meteorology* **35**: 165–185.

Mendel, J.M., 2001, *Uncertain Rule-Based Fuzzy Logic Systems: Introduction and New Directions*. Prentice-Hall, Upper Saddle River, NJ.

Mendel, J.M., and John R.I., 2002, Type 2 fuzzy sets made simple. *IEEE Transactions on Fuzzy Systems* **10**: 117–127.

Millington, A.C., 1996, Mapa de Comunidades Vegetales de la Estacion Biologica del Beni – Reserva de la Biosphera. Trinidad, Bolivia, Primer Congreso Internacional.

Okeke, F., and Karnieli, A., 2006, Linear mixture model approach for selecting fuzzy exponent value in fuzzy *c*-means algorithm, *Ecological Informatics* **1**: 117–124.

Robinson, V.B., 1988, Some implications of fuzzy set theory applied to geographic databases, *Computers Environment and Urban Systems* **12**: 89–98.

Robinson, V.B., and Strahler, A.H., 1984, Issues in designing geographic information systems under conditions of inexactness, *Proceedings of the 10th International Symposium on Machine Processing of Remotely Sensed Data*, Purdue University, Lafayette, pp. 198–204.

Robinson, V.B., and Thongs, D., 1986, Fuzzy set theory applied to the mixed pixel problem of multispectral landcover databases, in: B. Opitz (ed.), *Geographic Information Systems in Government*, A. Deerpak Publishing, Hampton, pp. 871–885.

Russell, B., 1923, Vagueness, *Australian Journal of Philosophy* **1**: 84–92.

Sorensen, R.A., 1985, An argument for the vagueness of the 'vague', *Analysis* **45**: 134–137.

Varzi, A.C., 2003, Higher-Order Vagueness and the Vagueness of 'Vague', *Mind* **112**: 295–298.

Verstraete, J., de Tré, G., de Caluwe, R., and, Hallez, A., 2005, Field based method for the modelling of fuzzy spatial data, in: F. Petry, V. Robinson, and M. Cobb (eds.), *Fuzzy Modeling with Spatial Information for Geographic Problems*, Springer, New York, pp. 41–69.

Williamson, T., 1994, *Vagueness*, Routledge, London.
Zadeh, L.A., 1965, Fuzzy sets, *Information and Control* **8**: 338–353.
Zadeh, L.A., 1975, The concept of a linguistic variable and its application to approximate reasoning – 1. *Information Sciences* **8**: 199–249.

INDEXING IMPLEMENTATION FOR VAGUE SPATIAL REGIONS WITH R-TREES AND GRID FILES

FREDERICK E. PETRY, ROY LADNER
Naval Research Laboratory, Stennis SpaceCenter, MS 39529, USA

MARIA SOMODEVILLA
Benemérita Universidad Autónoma de Puebla, Mexico

Abstract. We consider approaches to modeling spatial uncertainty with vague minimum bounding rectangles. Then two indexing approaches using grid files and R-trees are developed. Finally a number of spatial queries are illustrated based on these indexing schemes.

Keywords: minimum bounding rectangles, fuzzy sets, spatial queries, R-trees, grid files, vague spatial regions

1. Introduction

A vague region is one whose boundaries cannot be precisely defined. For this presentation of these types of regions, we separate them in their two main components: the core and the boundary. The core and the boundary are approximated by their minimum bounding rectangle (MBR) respectively. A fuzzy representation, called Fuzzy Minimum Bounding Rectangles (FMBR) (Somodevilla and Petry, 2003), has been developed in order to represent the different degrees of membership of the point located inside the vague region.

Querying vague regions is another important issue that is an open research line in the area of geographic information science. Here we describe alternative mechanisms to query vague regions having an FMBR representation and a more general MBR representation. The approaches taken under consideration are extensions of the grid file and R-trees spatial access methods. In particular we show how spatial queries by features and by location can be answered using the proposed query mechanisms.

So our goal is modeling and for indexing implementations the type of imprecision characterizing classes that for various reasons cannot have, or do not have sharply defined boundaries. To achieve this we consider the

A. Morris and S. Kokhan (eds.), Geographic Uncertainty in Environmental Security, 187–199.

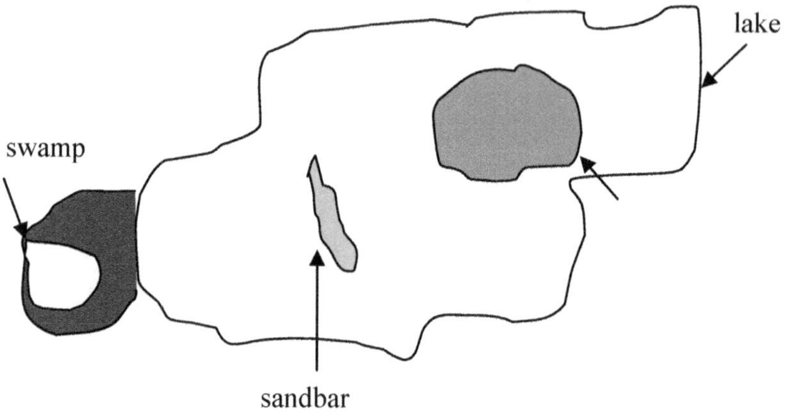

Figure 1. Example of vague spatial boundaries

integration of MBRs in an framework that captures all the semantics of the vague geographic phenomena or entities, incomplete boundaries, conflicting representations of the features, changing representation of dynamic phenomena and even imprecise observation.

Some of these are illustrated in Figure 1.

2. Related Work

Access methods have been classified as PAMs (Point Access Methods) and Spatial Access Methods (SAMs) (Ahn et al., 2001). The PAMs methods basically include: Grid File, kd-tree based (LSD-, hB- trees), Z-ordering and B+-tree. SAMs are used to retrieve information about spatial objects. They are based on the approximation of a complex spatial object by the minimum bounding rectangle (MBR) since MBRs preserve the most essential geometric properties of objects. SAMs include R-tree variations: R*-tree, R^+-tree and Hilbert R-tree.

3. Fuzzy Minimum Bounding Rectangle (FMBR)

Geographic features are a direct representation of geographic entities rather than geometric elements such as a point, line or polygon. A feature is then defined as an entity with common attributes and relationships. The FMBR (Somodevilla and Petry, 2003), represents the generalization of the underlying irregular polygon delimiting the fuzzy region since the FMBR encloses all the points of the map space where our feature of interest is located.

The FMBR can be also considered as the circumscribed rectangle (CR) of the underlying fuzzy polygon. Iterative generation of inner bounding rectangles is performed until we have the inscribed rectangle (IR) of the underlying object. So, the IR is the maximum inner rectangle inside the object and it corresponds to the core of the fuzzy region. Distances between the IR and the FMBR are used to represent the fuzzy boundary.

A spatial membership function based on Euclidean distance will be used to determine the degree of belonging of a feature to the fuzzy set. Thus, features inside the IR or core will have degree of membership of 1. This degree will be gradually decreased while we move away from the core. Points located outside of the FMBR will have a membership degree of 0.

An FMBR is a natural representation for many commonly occurring spatial situations. The problems of identifying a spatial boundary have been under considerable attention for the GIS area (Burrough, 1996). For example consider photointerpreters who are trying to label a forest in an image. There is clearly a region (core) which all agree is the heart of the forest and merits the specific labeling. However, as the forest thins into meadows all around, there is no sharp boundary delimiting the forest area. Rather the density of the trees decreases gradually until there is just open meadow. It is just such a situation that we are trying to model by means of an FMBR.

A graphical representation of the FMBR, as described above, is shown in Figure 2. The underlying vague region Ậ is approximated by the FMBR (Ậ). This first approximation is also called the circumscribed rectangle (CR) of the fuzzy region. In other words, the FMBR or CR corresponds to the minimal rectangle with edges parallel to the *x*- and *y*-axes that optimally enclose the vague region Ậ.

αMBR-cuts allow us to make finer distinctions inside the fuzzy region since *αMBR-cuts* are individual crisp regions inside the FMBR Thus, we can think of a fuzzy structured region as an aggregation of crisp α-level regions. αMBRs start to be defined from the edge of the FMBR(Ậ) to the core of (Ậ). The more external the *αMBR-cut*, the lower the degree of

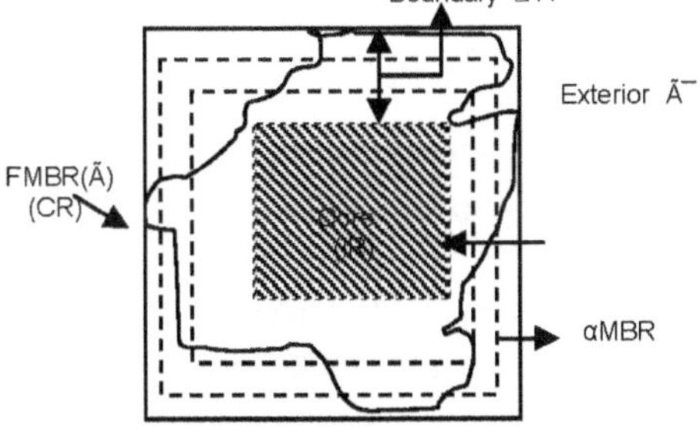

Figure 2. FMBR representation

membership in the fuzzy set representing (Ȃ) as locations which are closer to the core will have higher membership degrees. The shadowed rectangle labeled as *Core* corresponds to the inscribed rectangle. Since the *IR* is totally inside (Ȃ) we assume that the points in the core belong to the fuzzy region with a membership 1. Details about the representation and spatial relationships of FMBRs can be found in (Somodevilla and Petry, 2003),

4. Spatial Queries

A spatial database stores objects having spatial characteristics describing them. The spatial relationships among the objects are also important, and they are often required when querying the database. The main extensions that are needed for spatial databases are models that can interpret spatial characteristics. The basic extensions needed are two-dimensional geometric concepts, such as points, lines and polygons, in order to specify the spatial characteristics of the object. In addition, special indexing and storage structures are crucial to improve performance.

Spatial operations are required to operate on an object's spatial characteristics, for instance, to compute the distance between two objects, or to check whether two spatial objects overlap. The following categories represent the typical types of spatial queries:

4.1. FEATURE-BASED QUERY

Range query: It finds the objects of a particular type that are inside a given spatial area or within a particular distance from a given location, for example – select all rivers that pass within 3 km of a city? – or – are there any train depots within region X?

Spatial joins or overlays: These join objects of two types based on some spatial condition, such as the objects intersecting or overlapping spatially; for instance – find all cities that fall on major highways?

4.2. LOCATION-BASED QUERIES

Nearest neighbor query: It finds an object of a particular type that is closest to a given location, for example – what is the nearest waterway to train depot X?

An FMBRs database, besides the extensions needed to implement a spatial database has to manage the uncertainty associated with the borders of the geographic phenomena to be able to answer queries like: *find all locations that are about 500 m from a road* or *find all locations that are close to the road.*

5. Indexing and Storage Spatial Structures

For the type of spatial queries discussed to be answered efficiently, spatial techniques for spatial indexing are needed. Among the best-known techniques are the Grid Files and the R*-Tree. We now describe the use of Grid Files and R*-trees for storing and indexing vague regions represented by FMBRs.

5.1. GRID FILES

The grid file access method (Elmasri and Navathe, 2006), retrieves records by at most two disk access and efficiently handles range queries. This is done by using a grid of grid cells. All records in one grid cell are stored in the same bucket. However, several grid cells can share a bucket as long as the union of these grid cells forms a rectangle in the space of records. This guarantees that the records stored in the same bucket will be near to each other. Although the regions of the buckets are piecewise disjoint, together they span the space of records. To guarantee that data are always found with no more than two disk access for exact match queries, the grid itself is kept in main memory, represented by n-dimensional arrays called *scales*, where n is the number of search keys.

Since we are using FMBRs to represent vague regions, we have decided to store points belonging to the same αMBR in the same bucket to enhance querying performance. Thus, we might have many buckets as αMBRs. In Figure 3, the two buckets above the grid directory and the one at the bottom right corner store just one αMBR. On the other side, the bucket at the bottom left corner stores two different αMBRs. This does not violate the organization principle of the grid files because the points of a particular αMBR forms a rectangle. In addition, a relationship of closeness with respect to the core of

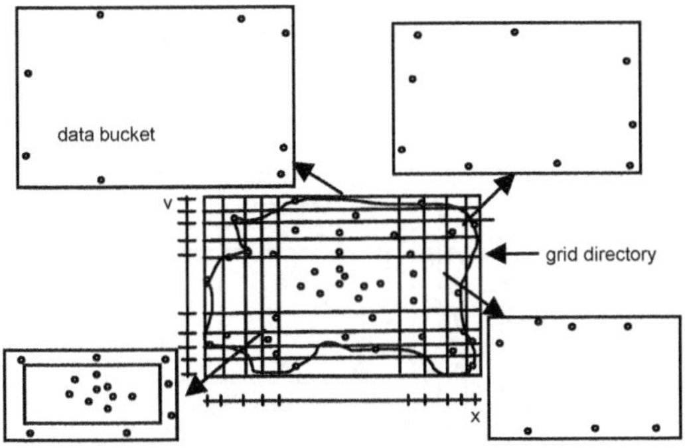

Figure 3. FMBRs grid file

the vague region is fulfilled. Moreover, points within two consecutive αMBRs are close enough among them and do not decrease the querying process performance, thus, they can be merged in the same bucket.

5.2. R-TREES

The R-tree (Guttman, 1984) is the basis of all R-tree variants. Each node corresponds to a disk page and an n-dimensional rectangle. Any entry in the tree is a pair (*ref*, *rect*), where *ref* is the address of the child node and *rect* is the MBR of all entries in that child node. The root has at least two children if not a leaf node. The number of entries in each node is between m (fill-factor) and M (number of entries that can fit in a node), where $2 \leq m \leq M/2$. All leaves are at the same level. Leaves contain entries of the same format, where *ref* points to a database object, and *rect* is the MBR of that object. An object appears in one, and only one of the tree leaves. R-trees are dynamic structures since insertion and deletion can be intermixed with queries and no periodic global reorganization is required. The external memory structure is multiway and it is indexed by MBRs.

R-trees present several weaknesses mainly due to the overlap between buckets regions at the same tree level. Moreover, the region perimeters should be minimized in order to avoid insertion problems. Insertion requires multiple paths of the tree, since the inserted spatial feature may intersect more than one intermediate node, and its clipping parts should be inserted in leaves under all such nodes. R*-trees are variations that avoid some of these problems. Representing FMBRs using an R*-tree structure was found very suitable since we can take advantage of the MBR representation of the objects in this model. Figures 4 and 5 correspond with our FMBR R*-tree description.

The representation of the FMBR and its respective αMBR-cuts are shown at the left upper corner of the Figure 4. Since we are interested in treating each αMBR-cut independently we have located each of them as

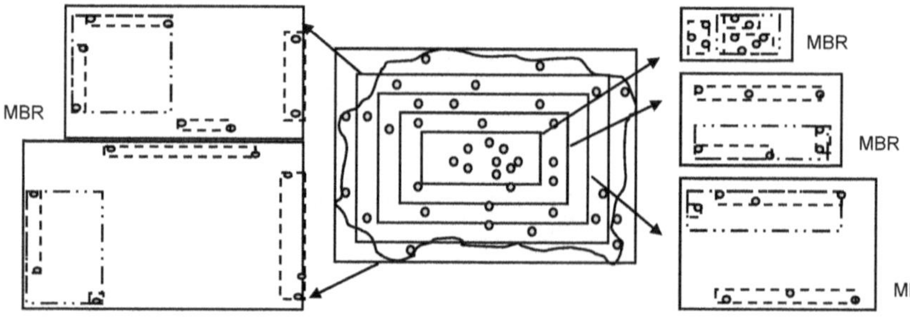

Figure 4. FMBR representation for R-tree

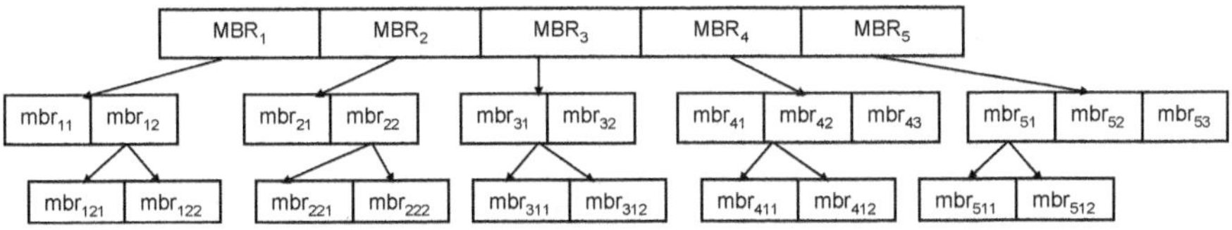

Figure 5. R-tree of Figure 4

root nodes of the tree. This structure allows us to access the features inside the vague region with a specific degree of membership following a unique path from the root. In addition, geographically close features belonging to the same αMBR-cut can be grouped in MBRs to improve the retrieval process.

The R*-Tree of the Figure 5 contains five nodes at the root corresponding to the core, and the four αMBRs approximating the boundary. The core αMBR_1 has two MBRS: mbr_{11} and mbr_{12}, and mbr_{12} contains mbr_{121} and mbr_{122}. A similar structure is maintained in the remaining nodes.

6. Querying FMBRs

In this section, we show the procedures to query FMBRs by location and by features. For each type of query we illustrate the use of Grid Files and R*-trees.

6.1. QUERYING FMBRS BY LOCATION

To answer queries by location is straightforward. Once we know the coordinates of a feature we can find the relevant cell by looking in the grid file. Consider again the query, – *what is the nearest waterway W_i to the train depot X?* –

6.1.1. *FMBR grid file*

The solution of the query is as follows:

Obtain the coordinates of the train depot X.
Use the coordinates to locate the bucket containing X.
Look for other data points representing waterways.

The searching starts in the same bucket where is X stored. If more than one waterway were found, the one having the least distance to X is selected for the answer. If no waterways were found in the same bucket as X, the

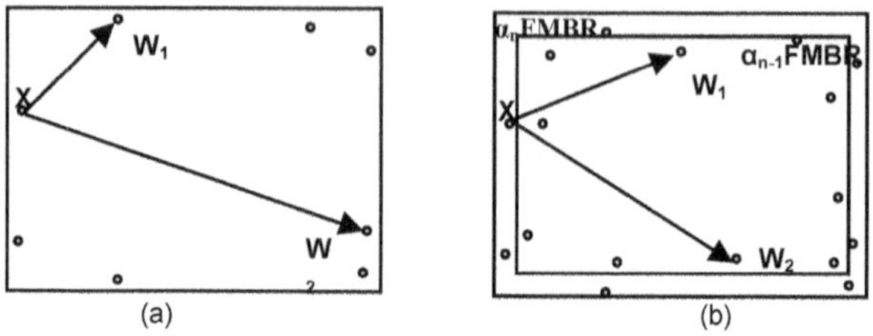

Figure 6. Querying by GFile location; searching of the nearest waterway to the train depot X

searching continues at the next inner αFMBR, otherwise we need to look for waterways at the next outer αFMBR and so on. Figure 6 shows an example of this kind of query.

In Figure 6, the αFMBR that contains X does not contain any waterways (a), so the searching continues in the next αFMBR. This αFMBR contains W_1 and W_2 waterways (b). Since the distance between the trains depot X and W_1 is the minimum, W_1 is the waterway that satisfy the answer.

6.1.2. *FMBR R*-tree*

The solution of the query is as follows:

Obtain the coordinates of the train depot X.
Use the coordinates to follow the path to the MBR containing X.

Since the train depot X is contained by the MBR_5 (see Figure 7), we used this root node to follow the path until the leaf corresponding to mbr_{511} which encloses X. Then, we searched in the contiguous MBR_4 and found W_1 and calculated its distance with respect to X. The search continues in the child node of mbr_{41} and another waterway is found. After the distance calculation with respect X, W_1 is found to correspond to the best solution.

6.2. QUERYING FMBRS BY FEATURES

In order to query the database by features we need to keep a look-up table mapping features and their respective locations. To answer the previous query – are there any waterway in the region X? – the steps below are taken.

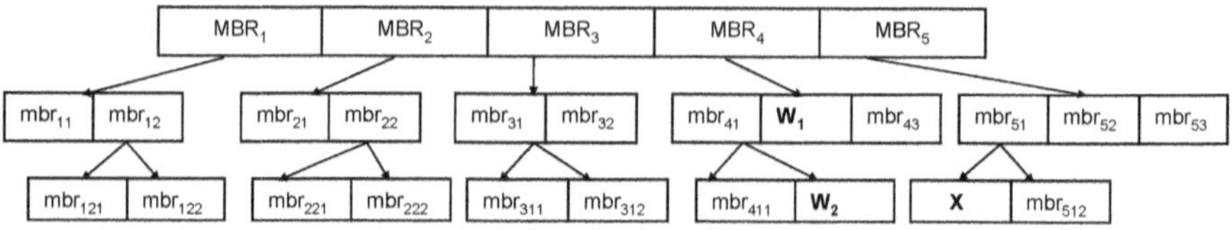

Figure 7. Query by R*-tree location; searching of the nearest waterway to the train depot X

6.2.1. *FMBR grid file*

Find the coordinates of the region X from the table. These coordinates should correspond to the corners of the FMBR approximating X.

Each αFMBR of the entire FMBR needs to be checked for any waterway lying inside of it. Since features within an αFMBR have a degree of membership for the fuzzy region depending on the distance to the FMBR's core, different degrees of membership will be assigned to the waterways belonging to the region X. An example of the evaluation of this query is shown in Figure 8. First the αFMBR enclosing the region X is located. Then the search for the waterways starts at the αFMBR closest to the core of the region (inner αFMBR). Thus, W_1 is the waterway that satisfies the query best, since its membership in the fuzzy region X is higher than the membership of W_2.

6.2.2. *FMBR R*-tree*

Obtain the coordinates of the region X from the table. These coordinates should correspond to the corners of the MBR_5.

Each αFMBR of the entire FMBR needs to be checked for any waterway lying in it. We start the search for the inner rectangle which has the higher membership. An example of evaluation of this query is shown in Figure 9.

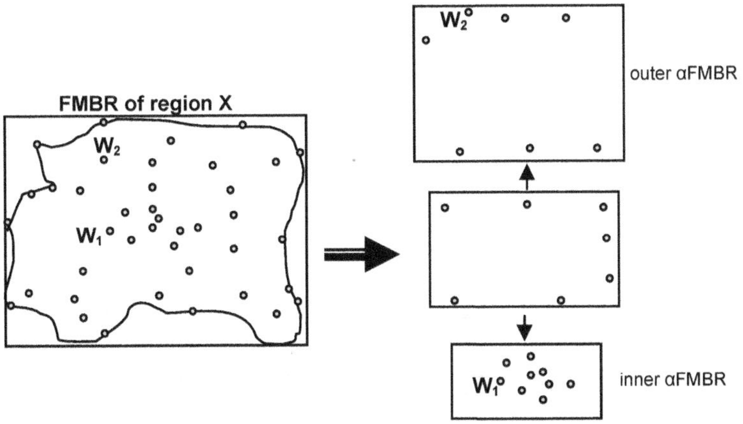

Figure 8. Query by grid file features; finding any waterway in region X

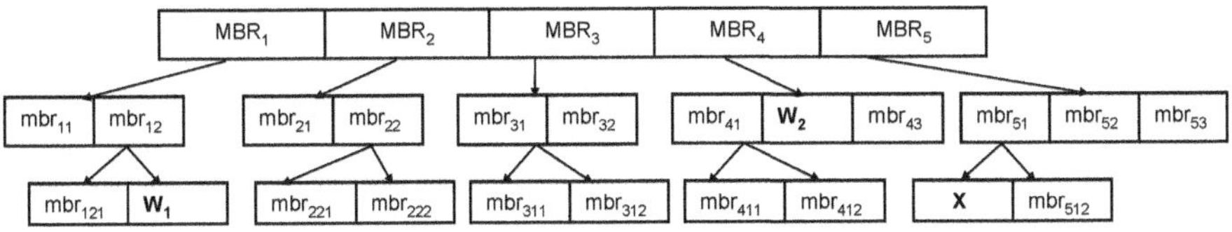

Figure 9. Query by R*-tree features; finding any waterway in the region X

From Figure 9, we found waterways in MBR_5 and MBR_1. Since MBR_1 encloses the core of the vague region, W_1 is the best answer for the query.

7. Possible and Certain Spatial Regions Indexing

In this section we discuss some approaches for modeling and indexing vague spatial regions. In this context a vague spatial region is one in which we are able to specify inner (certain) and outer (possible) region boundaries but for which the αMBR cuts used above are not suitable due to the inhomogeneous nature and discontinuities of the region.

To utilize MBRs for vague regions we define a vague MBR (VMBR) as consisting of nested rectangles. The inner rectangle (IVM) is the MBR over the core of the vague region (certain region or membership = 1). The outer rectangle (OVM) is an MBR over the outer (possible) boundary of the vague region.

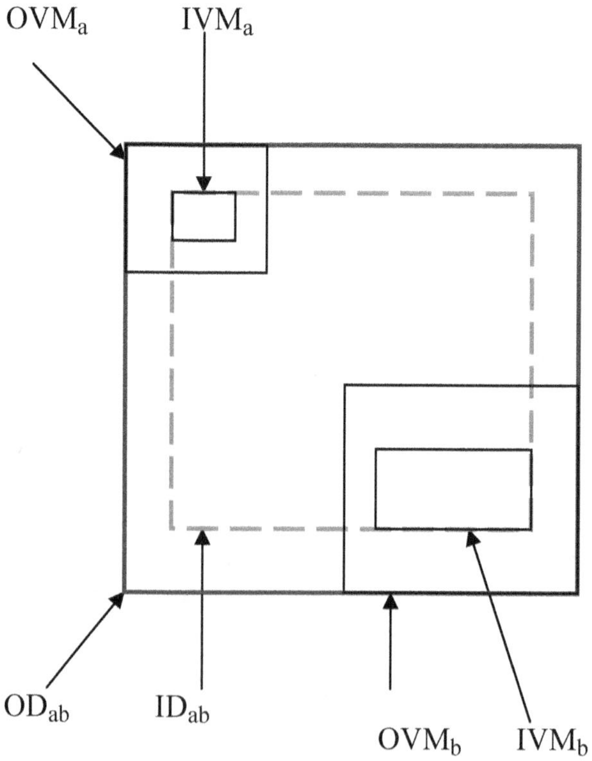

Figure 10. Directory rectangles for vague MBRs

This approach allows us to then consider common indexing approaches such grid files or R-trees. We consider in some detail alternatives for constructing R-trees for VMBRs and spatial queries using these structures. So now we wish to consider how to effectively index the VMBRs. As we have discussed for FMBRs, R-trees are a most important indexing approach used for spatial databases. Recall that the interior nodes of an R-tree are the directory rectangles (minimum bounding) surrounding the VMBRs.

Here we are proposing to use multiple directory rectangles to enclose the inner VMBRs (IVM) as well as the outer VMBRs (OVM). This is illustrated in the space decomposition shown in Figure 10. Here we have two vague spatial regions, A and B, with the corresponding VMBRs:

$$(OVM_a, IVM_a) \text{ and } (OVM_b, IVM_b)$$

The solid line is the outer directory rectangle (OD_{ab}) which is based on the two VMBRs outer rectangles, OVM_a and OVM_b. Similarly the dotted line corresponds to the inner directory rectangle ID_{ab} enclosing the the VMBRs' inner rectangles, IVM_a, IVM_b.

Corresponding to this will be an actual R-tree. Each R-tree node can store some maximum number (M) of rectangles and associated pointers based on the size of allowed disk buckets. For simplicity in the following discussion we will assume the maximum number M is 2. This yields an R-tree as shown in Figure 11.

In the structure are two directory rectangles at an interior node. The two leaf nodes below them will contain the VMBRs. However what is not clear is how the IVMs and OVMs should be stored in the R-tree to optimize accesses for querying.

We can consider two alternatives for these nodes. First we can have Design 1, for which the inner and outer VMBRs are in the same nodes: Node 1: [OVM_a, IVM_a] and Node 2 [OVM_b, IVM_b]. Another possible

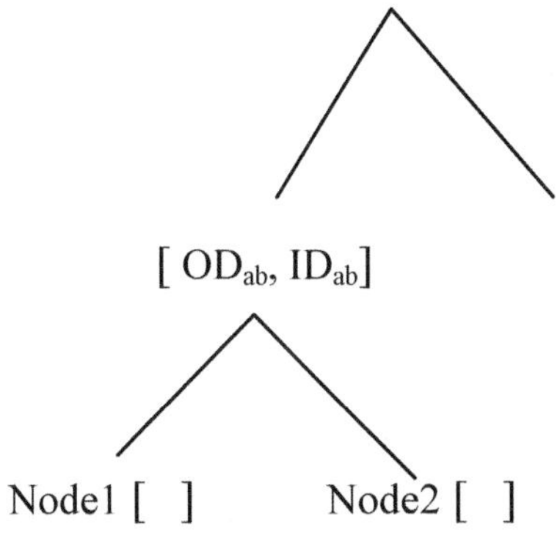

Figure 11. R-tree corresponding to Figure 10

strategy, Design 2, is to group the inner VMBRs and the outer VMBRs together for each directory rectangle node, Node 1: [OVM_a, OVM_a] and Node 2 [IVM_b, IVM_b].

To evaluate the relative merits of the two approaches we must consider the different potential queries cases. For point queries there are several cases and we illustrate representative ones. We consider two types of point queries, first an intersection type query, for example, "Does point p fall inside of Figure A?." The other is a classification type query. Also for each type of query we can reflect whether the result is just possible or is certain. We next show two cases of the first type of query and the number of accesses for each design.

Case 1: $p \in OD_{ab}$ and $\notin ID_{ab}$

Certain (p) No – already answered since $p \notin ID_{ab}$

Possible (p) if $p \notin OVM_a$ and $p \notin OVM_b$
Design 1 1 access
Design 2 2 accesses

Case2: $p \in OD_{ab}$ and $\in ID_{ab}$

Not Certain (p) if $p \notin IVM_a$ and $p \notin IVM_b$
Design 1 1 access
Design 2 2 accesses

Now we consider a classification question of the form, where is p?

Case: $p \in OVM_a$ and $\notin IVM_a$

Answer is: Possibly but not certainly OVM_a
Design 1 two accesses –

One access to check $p \in OVM_a$ then must access to check $p \notin IVM_a$
Design 2 one access to check both conditions

8. Summary

An FMBR is an extension of a determinate region model which facilitates the treatment of vague regions by applying fuzzy sets concepts. A vague region is modeled by using an FMBR. Querying an FMBR, is the focus of this paper. A representation based on a grid file indexing mechanism has been proposed. In addition R*-tree was proposed as other alternative mechanism to realize spatial queries. Both strategies allow us to query the FMBRs by location and features. We found the R*-Trees FMBR more suitable in solving queries based on location and features. The reason for this given by the structure based on concentric MBRS of the FMBRs. In representing vague regions, we are interested in the degrees of membership of the different features lying in the region and R*-trees allow us to make this distinction in a very natural and efficient way.

Acknowledgments

The authors would like to thank the Naval Research Laboratory's Base Program, Program Element No. 0602435N for sponsoring this research.

References

Ahn, H., Mamoulis, N., and Wong, H., 2001, Survey on multidimensional access methods, University of Utrecht Technical Report, UU-CS-2001-14, www.cs.uu.nl/research/techreps/.

Burrough, P., 1996, Natural objects with indeterminate boundaries, *Geographic Objects with Indeterminate Boundaries*, Taylor and Francis, London, pp. 3–28.

Elmasri, R., and Navathe, S., 2006, *Fundamentals of Database Systems*, Fifth Edition, Addison Wesley.

Guttman, A., 1984, R-trees: a dynamic index structure for spatial searching, *Proceedings ACM SIGMOD International Conference on Management of Data*, pp. 47–57.

Somodevilla, M., and Petry, F., 2003, Approximation of topological relations on fuzzy regions: an approach using minimal bounding rectangles, *Proceedings of the NAFIPS03 Conference*, Chicago, Il, pp. 371–376.

ASSOCIATION RULE MINING USING FUZZY SPATIAL DATA CUBES

NARIN IŞIK, ADNAN YAZICI
Middle East Technical University, Department of Computer Engineering

Abstract. The popularity of spatial databases increases since the amount of the spatial data that need to be handled has increased by the use of digital maps, images from satellites, video cameras, medical equipment, sensor networks, etc. Spatial data are difficult to examine and extract interesting knowledge; hence, applications that assist decision-making about spatial data like weather forecasting, traffic supervision, mobile communication, etc. have been introduced. In this thesis, more natural and precise knowledge from spatial data is generated by construction of fuzzy spatial data cube and extraction of fuzzy association rules from it in order to improve decision-making about spatial data. This involves an extensive research about spatial knowledge discovery and how fuzzy logic can be used to develop it. It is stated that incorporating fuzzy logic to spatial data cube construction necessitates a new method for aggregation of fuzzy spatial data. We illustrate how this method also enhances the meaning of fuzzy spatial generalization rules and fuzzy association rules with a case study about weather pattern searching. This study contributes to spatial knowledge discovery by generating more understandable and interesting knowledge from spatial data by extending spatial generalization with fuzzy memberships, extending the spatial aggregation in spatial data cube construction by utilizing weighted measures, and generating fuzzy association rules from the constructed fuzzy spatial data cube.

Keywords: fuzzy spatial data cube, spatial data cube, fuzzy data cube, fuzzy association rules, spatial knowledge discovery

1. Introduction

Decision support systems (DSSs) are database management systems that run queries efficiently to understand the trends, make decisions and predictions. They necessitate historical data that are consolidated from

A. Morris and S. Kokhan (eds.), Geographic Uncertainty in Environmental Security, 201–224.

heterogeneous sources by data warehouses. Data warehouses support information processing by providing a solid platform of consolidated, historical data for analysis. Numerical data is stored in multidimensional fashion and analyzed by OLAP operations through high-level aggregates that summarize the numeric values. A data warehouse is based on a multidimensional data model that views data in the form of a data cube while relational model views data in the form of tables. A data cube allows data to be modeled and viewed in multiple dimensions.

In "Fuzzy OLAPs," OLAP mining and fuzzy data mining are combined to get benefit of the fact that fuzzy set theory treats numerical values in more natural way, increases the understandability, and extracts more general rules since numerical data are interpreted with words (Laurent and Bouchon-Meunier, 2000, 2001; Laurent, 2001). Fuzzy OLAPs are performed on fuzzy multidimensional databases in which multidimensional data model of data warehouses is extended to manage imperfect and imprecise data (i.e., bad sales) from real world and run more flexible queries (i.e., select middle sales).

Furthermore, Kuok et al. (Kuok, 1998), has generated fuzzy association rules by introducing significance and certainty factors and proposing methods for computing these factors.

Spatial databases are able to store information about the position of individual objects in space. On the other hand, many applications that assist decision-making about spatial data like weather forecasting, traffic supervision, or mobile communications necessitate summarized data (i.e., general weather patterns for regions, number of cars in an area, phones serviced by a cell). Moreover, creation of maps from satellite images and usage of telemetry systems, remote sensing systems or medical imaging results in a huge amount of spatial data. That causes to difficulties when examining large amounts of spatial data and extracting interesting knowledge or characteristic rules from them. Obtaining this information from operational (i.e., transactional) spatial databases is quite expensive. In that point, spatial data warehouses and OLAP becomes crucial for spatial knowledge discovery. Stefanovic et al. (Stefanovic, 2000) has studied methods for spatial OLAP by combining nonspatial OLAP methods and spatial databases. They propose a model for spatial data warehouses, which has both spatial and nonspatial dimensions and measures in the form of regions in space; and a method for spatial data cube construction called "object-based selective materialization."

In this study, spatial data cubes and fuzzy data cubes are tried to be harmonized in order to get benefit from the strengths of both of these concepts. Spatial and nonspatial dimensions and measures considered for spatial OLAP in (Stefanovic, 2000), spatial generalization described in

(Lui and Han, 1993) and fuzzy association rules asserted in (Kuok, 1998) are combined in one study and further improved. Better analysis and understanding of huge spatial data by using fuzzy set theory for spatial data cubes are aimed. More specifically, spatial generalization algorithms, that were previously mentioned (Lui and Han, 1993), are enhanced by introducing fuzzy logic in determining high-level concepts (i.e., membership values of high-level concepts are calculated). Moreover, a new aggregation method is introduced for fuzzy spatial data cube in which membership values of the more significant regions have greater weight for the aggregated region. Furthermore, we illustrate how this method is used for fuzzy spatial generalization rules and fuzzy association rules. Fuzzy association rule mining is applied over the generated fuzzy spatial data cube and computation of significance and certainty factors are adapted according to the cube, instead of applying spatial association rules (Han and Kamber, 2001) which have more complex computation.

This study differentiates from the previous studies about spatial knowledge discovery in the following ways:

Generalization algorithms in (Lui and Han, 1993) are extended by fuzzy high-level concepts and their memberships. Hence, more understandable and meaningful generalization rules are extracted due to fuzzy set theory (i.e., generalizations like "hot region with 89% reliability" instead of "[20–25]°C temperature region").

Spatial aggregation in spatial data cube construction is extended by calculating fuzzy memberships of dimensions and measures for the aggregated cells considering more significant regions with greater weight. More accurate generalization rules are extracted at higher levels of abstraction.

Deviations in precision values for reliability of characteristics of spatial data along time can be tracked.

Each cell in a fuzzy spatial data cube has its own membership for the values of fuzzy dimensions and fuzzy measures it satisfies. On the contrary, in fuzzy data cube, cells in a slice (i.e., cells that have a common value for one dimension) of a fuzzy data cube has the same membership.

Fuzzy hierarchies are handled during fuzzy spatial data cube in order to increase the level of abstraction (i.e., level of precise knowledge extracted from spatial data).

In previous studies, spatial association rules were mined from spatial data which has a high computational complexity. In the constructed fuzzy spatial data cube, more flexible association rules can be discovered by mining fuzzy association rules and avoiding the complexity of spatial association rules since data are both fuzzy and spatial.

In the rest of the thesis, first, background is given in Chapter 2 that includes fuzzy OLAP, spatial OLAP, and data mining. The methodology followed in fuzzy spatial data cube construction and its application in fuzzy association rule mining is explained in Chapter 3. The case study "Weather Pattern Searching" is introduced in Chapter 4. Finally, Chapter 5 includes the final comments and Chapter 6 draws a number of conclusions that summarizes the study.

2. Related Work

Spatial data cubes have improved the quality of decision-making about huge sizes of spatial data. On the other hand, fuzzy data cubes have enabled the extraction of relevant knowledge in a more human understanding way and with a certain precision about the reliability of that knowledge. Interesting data can be extracted from both spatial data cube and fuzzy data cube by association rules.

2.1. FUZZY OLAP

The need to handle imperfect data that are either uncertain or imprecise and run flexible queries on warehouses has motivated studies on fuzzy OLAP. Fuzzy OLAP enables the extraction of relevant knowledge in a more natural language to increase the understandability and gives results to the queries with a certain precision about the reliability of the knowledge (Laurent and Bouchon-Meunier, 2000; Laurent, 2001). Fuzzy summaries are generated from fuzzy multidimensional databases like "Most sales are medium: truth value = 0.16" in (Laurent and Bouchon-Meunier, 2001; Laurent, 2001). Fuzzy summary generation is based on algorithms for association rules.

In the fuzzy data cube displayed in Figure 1, the domain of the dimension "product" can be defined as $\text{Domain}_{\text{Product}}$ = {(oven,0.7), (television,1), (refrigerator,0.8)}. That means slices of the cube belong to the cube at some extent as the slice corresponding to "oven" along production dimension belongs to the cube with degree 0.7. Additionally, each cell in the fuzzy cube, belongs to the cube with a degree, that is, confidence value. The cube is represented as Product×Location×Time = Sales×[0,1]. The measure values also have fuzzy labels and membership values in each cell.

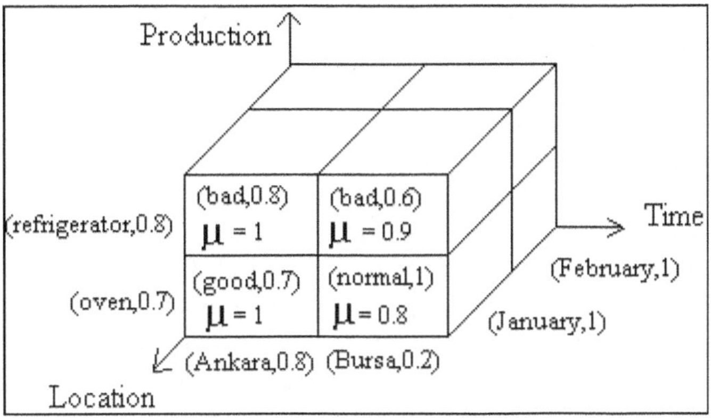

Figure 1. Fuzzy data cube

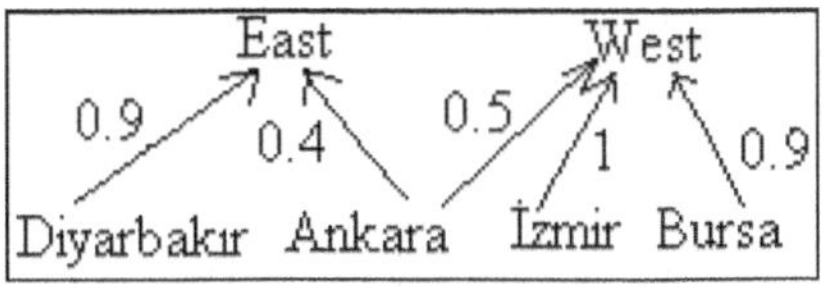

Figure 2. Fuzzy hierarchies

Fuzzy hierarchies are represented to indicate that some values may gradually belong to more than one of the defined higher levels like in Figure 2 (Laurent and Bouchon-Meunier, 2001; Laurent, 2001).

2.2. SPATIAL OLAP

Spatial OLAP provides efficient spatial OLAP operations for the summarization and characterization of large sets of spatial objects in different dimensions and at different levels of abstraction. Spatial objects have fast and flexible representation for their collective, aggregated, and general properties.

Knowledge discovery in spatial databases corresponds to the extraction of interesting spatial patterns and features, general relationships between spatial and nonspatial data, and other implicit general data characteristics (Lui and Han, 1993; Zaiane, 2002). A generalization-based knowledge discovery mechanism is developed to integrate attribute-oriented induction on non-spatial data and spatial merge and generalization of spatial data in (Lui and Han, 1993). Generalization rules, that is, general data characteristics and/or relationships, are extracted. Induction is done via ascending the thematic concept hierarchies and spatial hierarchies, higher levels (Lui and Han, 1993; Zaiane, 2002; Stefanovic, 1997). For example, suppose that a hierarchy is defined as "corn," "wheat," and "rice" being the leaves and "grain" being their parent, regions that grow corn, wheat, and rice can be generalized as "grain-production-area." Moreover, regions with precipitation measurements between 2.0 and 5.0 can be generalized as "wet-area."

Stefanovic et al. (Stefanovic, 2000) propose a model for spatial data warehouses that has both spatial and nonspatial dimensions and measures and a method for spatial data cube construction called object-based selective materialization. A spatial data cube may be used to view the weather patterns on a map by region, month, and different combinations of temperature and precipitation like "hot and wet regions in the Marmara Region." Spatial objects are summarized and characterized in different dimensions and at different level of abstraction by spatial OLAP operations in spatial data cubes. Hierarchies can be defined for dimensions as in Figure 3 and data can be summarized according to them as in Figure 4.

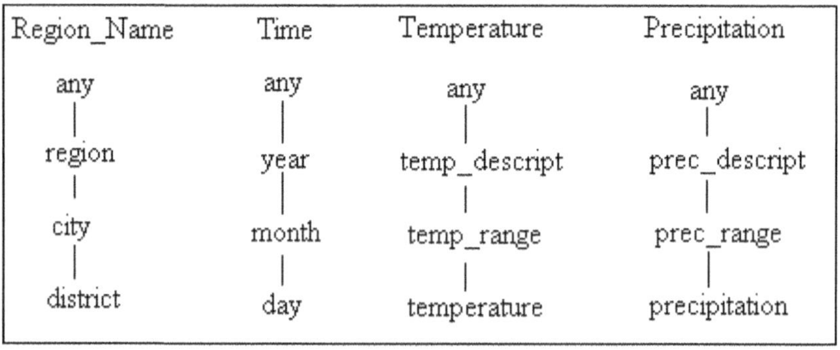

Figure 3. Hierarchies for the star schema

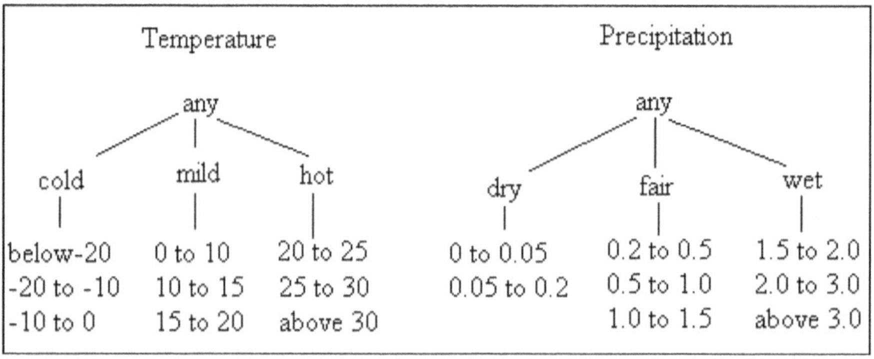

Figure 4. Example data for hierarchies in star schema

The result obtained when a roll-up operation is applied by the use of the hierarchies is illustrated in Table 1:

TABLE 1. Result of a roll-up operation

Precipitation	Region map
dry	{AK04, AK07, …, VS67}
fair	{AG10, AG05, …, TP90}
…	…

Different methods for computing spatial data cubes like online aggregation, materialization of all/some/parts of the cuboids, materialization of cuboids roughly, combining spatial indexing with preaggregated results, multiresolution amalgamation exist (Han and Kamber, 2001; Stefanovic, 2000; Harinarayan, 1996; Zhou et al., 2002; Prasher, 2004; Papadias, 2001; Zhou, 1999; Bedard, 2001).

2.3. ASSOCIATION RULES

Association rule mining (Han and Kamber, 2001; Zaiane, 2002; Kelkar, 2001; Kuok, 1998) is one of the functionalities of data mining and corresponds to finding frequent patterns, associations, correlations, and causal structures among sets of items or objects in information repositories.

It has application areas in basket data analysis, cross marketing, catalog design, loss-leader analysis, clustering, classification, etc.

Association rule mining was first aimed at discovering associations between items in transactional databases. Given D = $\{T_1 \ldots T_n\}$ a set of transactions and I = $\{i_1 \ldots i_n\}$ a set of items such that any T_i in D is a set of items in I. An association rule is an implication A->B where A and B are subsets of T_i given some support and confidence thresholds. In other words, an association rule is a rule that correlates the presence of one set of items with that of another set of items. The rule has the form "Body->Head [support, confidence]" (or "If X is A then Y is B") where support is the probability that a transaction contains the Body and confidence is the conditional probability that a transaction having the Body also has the Head.

Support of "Body->Head" = (# of transactions containing Body)
(total # of transactions)

Confidence of "Body->Head" = (# of transactions containing both Body and Head)
(# of transactions containing Body)

Association on a data warehouse requires aggregation to be performed at different levels, which can be slow. Since an OLAP cube has precomputed results (i.e., count of tuples), performing association on an OLAP cube is much faster.

In order to find associations, first, all frequent items are found. Frequent items correspond to items that are more frequent, that is, have supports greater than the initially determined minimum support. Then, frequent items are combined into item sets. After all item sets are found, they are used to produce association rules according to the initially defined minimum confidence. While generating association rules, the most important thing is to find the frequent item sets. The most famous algorithm is the Apriori algorithm that has many variations and improvements on it. One problem with the Apriori algorithm is that it misses all item sets with recurrent items.

Spatial association rules also have the form "Body->Head [support, confidence]." Here, "Body" and "Head" can be sets of spatial or nonspatial predicates such as (Han and Kamber, 2001) topological relations (intersects, overlaps, disjoint, etc.), spatial orientations (left_of, west_of, under, etc.), and distance information (close_to, within_distance, etc.), i.e., Is_a(x, large_town) ^ intersect(x, highway) -> adjacent_to(x, water) [7%, 85%]. Spatial associations are mined in two steps. In the first step, rough spatial computation is done to filter out the irrelevant spatial objects. MBR or R-tree is used for rough estimation. In the second step, detailed spatial algorithm is applied to refine the mined rules. Algorithm is applied only to the objects that have passed the rough spatial association test with a value greater than the minimum support (Han and Kamber, 2001).

TABLE 2. Record containing membership values

	Temperature		Precipitation		Area	
	Label	Membership	Label	Membership	Label	Membership
t_1	Hot	0.9	Dry	0.2	Large	0.5
t_2	Hot	0.7	dry	0.4	Large	0.8
t_3	Hot	0.8	Dry	0.3	Large	0.2
...	...	...	...	...	...	...

Similarly, fuzzy association rules have the form "If X is A then Y is B" where X and Y are disjoint sets of attributes and A and B are fuzzy sets that describe X and Y respectively (Kuok, 1998). Fuzzy association rules are more understandable to human because fuzzy sets handle numerical data better since they soften the sharp boundaries of data. The semantics of the rule is when "X is A" is satisfied it can be implied that the consequent part "Y is B" is also satisfied. Interesting rules have enough significance and high certainty factors. Significance is for the satisfiability of the item sets and certainty is for the satisfiability of the rules.

While generating fuzzy association rules, first large item-sets, those with significance higher than a user-specified threshold, are found. The significance is calculated by summing the votes of all records for the specified item-set and by dividing that sum to the count of the records. A vote of the record corresponds to the production of the membership values for the fuzzy sets in A that are described for X. For example, if X = {temperature, precipitation} and A = {hot, dry}. Suppose that we have the following records T = $\{t_1, t_2, t_3 \ldots\}$ depicted in Table 2. Significance of the rule is $S_{(X,A)} = (0.9 \times 0.2 + 0.7 \times 0.4 + 0.8 \times 0.3)/3 = 0.23$.

After discovering large item-sets, interesting rules are generated according to the certainty factor that is computed as $C((X,A),(Y,B)) = S(Z,C)/S(X,A)$ where $X \subset Z$, $Y = Z - X$ and $A \subset C$, $B = C - A$. For the example above, if Y = {area} and B = {large}, $C_{((X,A),(Y,B))} = ((0.9 \times 0.2 \times 0.5 + 0.7 \times 0.4 \times 0.8 + 0.8 \times 0.3 \times 0.2)/3)/0.23 = 0.517$.

In short, if the rule has enough significance it is determined as one of the large item-sets. Then, if it also has enough certainty it is specified as one of the interesting association rules in the database.

3. Fuzzy Spatial Data Cube

There is a huge amount of spatial data available which are not useful unless knowledge is obtained from them. Especially, GISs need to store, manipulate, and analyze voluminous amounts of spatial data timely and conclude specific decisions. Spatial data warehouses are very suitable for systems that need to store large amounts of spatial data and analyze them like GIS since spatial data warehouses integrate different sources together, enable characterization of spatial data, summarize data in different dimensions at different levels of abstraction, and facilitate discovery of knowledge and decision-making.

Equally important, fuzzy data warehouses incorporate fuzzy logic into their multidimensional data model and construct fuzzy data cubes in order to increase the understandability and nature of the extracted knowledge. They also give results to queries with a certain precision about the reliability of that knowledge.

In this study, more understandable and precise knowledge is generated from spatial data by construction of fuzzy spatial data cube and extraction of fuzzy association rules from the corresponding cube. In the following subchapters, it is illustrated how spatial data cubes and fuzzy data cubes can be harmonized together and how this benefits to spatial knowledge discovery.

3.1. FUZZY SPATIAL DATA CUBE CONSTRUCTION

Data mining discovers nontrivial and interesting knowledge or patterns from data. It has functionalities like characterization, comparison, classification, association, prediction, cluster analysis, and time-series analysis. In this study, characterization and association aspects are considered over fuzzy spatial data cube to discover precise multilevel knowledge from spatial data.

Characterization (i.e., generalization) can be used to generalize task-relevant data into generalized data cube. Characteristic rules, which are extracted from a generalized data cube, summarize general characteristic of user-specified data. Similarly, characteristic rules, which are extracted from a fuzzy spatial data cube, can summarize the climate data for a region with the extension that they can also present the precision. The raw data for one region can be generalized into concepts like cold (0.9), mild (0.7) and hot (0.5) for temperature, and dry (0.28), wet (0.72) for precipitation with the precision values that indicates the degree of reliability of the generalization. Subregions that are described by the same high-level concepts can be aggregated together with a recomputed precision which is the subject of this thesis.

Instead of generalizing spatial data to "[20–25] temperature regions," "[25–30] temperature regions" and "[above 30] temperature regions" and then aggregating them to "hot regions"; generalizing each spatial datum to "hot region" with the precision value μ_{hot} for the reliability to that generalization and then aggregating them to "hot regions" with a new μ_{hot} for the aggregated regions is more meaningful and natural. Different temperature values will cause to generalizations with different precision values. Introducing fuzzy logic to spatial generalizations helps to have more smooth generalizations (Figure 5).

Fuzzy spatial data cube considered in this study combines both some features of spatial data cube and fuzzy data cube that take place in the literature and also differentiates from them at some points. These similarities and differences are explained in more details below:

In fuzzy spatial generalization rules, besides the spatial generalization (i.e., hot), the membership value of the generalization is also computed (i.e., hot (0.96)) according to the defined fuzzy labels and membership functions (Table 3).

In contrast to spatial data cube construction proposed by Stefanovic (Stefanovic, 2000) in which numeric values are first generalized to ranges and then to more descriptive names, in fuzzy spatial data cube they are generalized directly to their descriptive names (fuzzy labels) with their membership values.

Fuzzy spatial data cubes are very similar to the spatial data cube considering their dimensions and measures with the exception that fuzzy generalization rules are discovered at multiple levels of abstraction from spatial data by the help of defined fuzzy hierarchies.

In fuzzy spatial data cubes, measures are also fuzzified (i.e., generalized with their precision values) as in fuzzy data cubes.

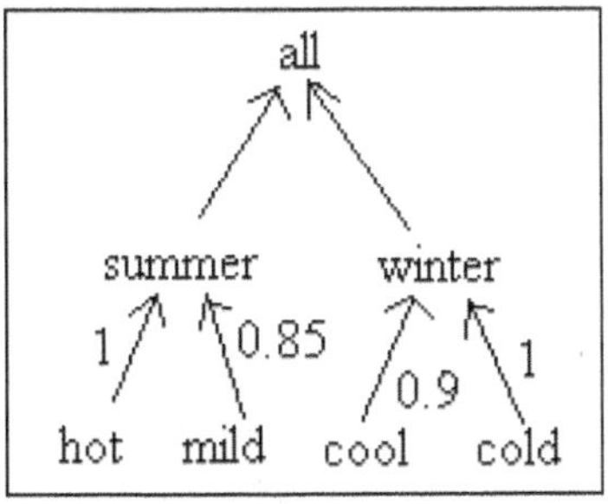

Figure 5. The season fuzzy hierarchy for the temperature dimension

TABLE 3. Generalized geo-objects

Temperature	$\mu_{temperature}$	Season	μ_{season}	Precipitation	$\mu_{precipitation}$	Area	ids
Mild	0.99	Summer	0.84	Wet	0.93	30	R1
Mild	0.98	Summer	0.83	Wet	0.99	50	R2
Hot	0.99	Summer	0.99	Wet	0.89	65	R3
Hot	0.99	Summer	0.99	Wet	0.88	38	R4
Hot	0.64	Summer	0.64	Dry	1.0	70	R5
Cool	0.97	Winter	0.87	Wet	1.0	43	R6

TABLE 4. Spatial generalization

Region	Temperature
R1	Hot
R2	Hot
R3	Hot
R1, R2, R3	Hot

TABLE 5. Fuzzy spatial generalization

Region	Temp	Temperature membership	Area
R1	Hot	0.98	40
R2	Hot	0.67	30
R3	Hot	0.56	50
R1, R2, R3	Hot	$(0.98 \times 40 + 0.67 \times 30 + 0.56 \times 50) / 120 = 0.73$	120

Generalization rules for spatial data are extended for fuzzy spatial data cube since generalization is done not only according to the determined characteristic but also considering both the characteristic and membership values of the generalized data. A new membership is needed to be calculated for the generalized value in the fuzzy spatial data cube for the aggregated regions as illustrated in Table 4 and Table 5. The membership value of the generalized aggregated region is computed by considering the weight of one of the measures according to the context of the application.

In Laurent's study (Laurent et al., 2000; Laurent and Bouchon-Meunier, 2001; Laurent, 2001) for fuzzy cubes, each slice corresponds to the cube with a membership value, that is, a value of one dimension has the same membership value for all the cells in the slice. But in our fuzzy spatial data cube each cell has its individual membership value for the corresponding dimension value since spatial objects might have common properties but each spatial object might have that property with a different degree than other spatial objects as displayed in Figure 6.

In Laurent's study (Laurent et al., 2000; Laurent and Bouchon-Meunier, 2001; Laurent, 2001) aggregation is done by computing the degree to which the aggregated cell belongs to the cube, as explained in section 2.3 "Fuzzy OLAP." That is done by computing the arithmetic cummilation of the membership values of cells that are being aggregated. On the contrary in our study, aggregation is done by multiplying membership values of each cell with the weighted value of the tuple, summing these multiplications and dividing to the sum of the weighted values.

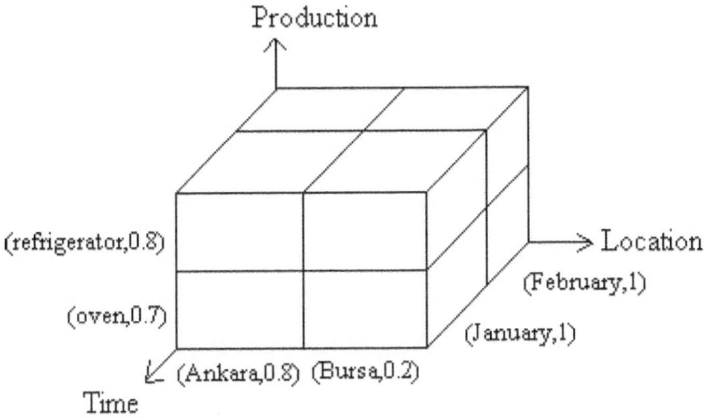

a) Fuzzy Data Cube

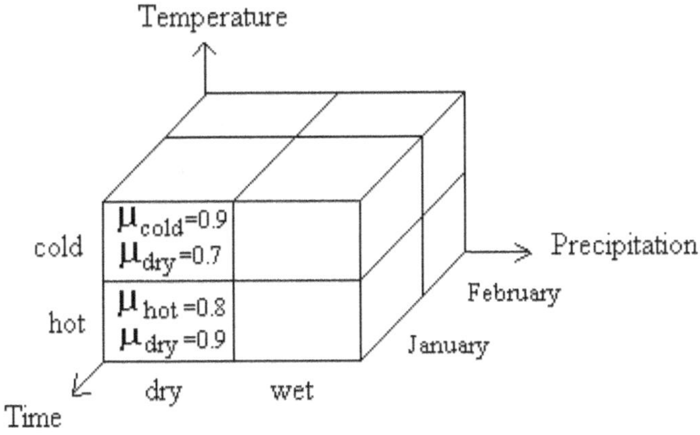

b) Fuzzy Spatial Data Cube

Figure 6. Dimensions and their memberships

With the use of fuzzy spatial data cubes, generalization rules such as "Ankara was %80 hot and %78 dry in June, 2003." can be easily extracted. Moreover, an additional generalization rule like "Ankara was %85 hot and %96 dry in June, 2004" could help to conclude that the increase in hotness for the region Ankara has also increased dryness of weather, in other words, increase in temperature has decreased the precipitation in one-year period. Incorporating fuzzy logic in spatial data cubes increases the reliability to the generalizations due to the computed precisions and helps to identify the deviations in properties of spatial regions through time.

The construction of fuzzy spatial data cube constitutes of the following steps (Figure 7):

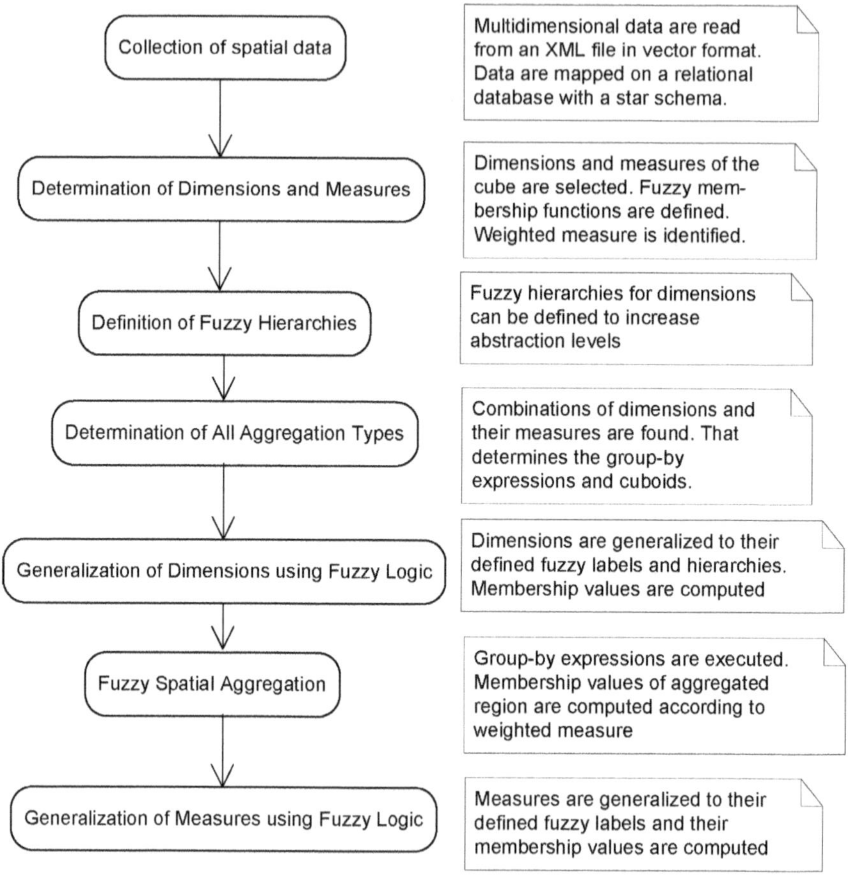

Figure 7. Fuzzy spatial data cube construction steps

3.2. MEANING OF VALUES IN THE FUZZY SPATIAL DATA CUBE

The crisp data that were available at the beginning are generalized and fuzzified. The meaning of the obtained knowledge can be commented as in Table 6.

In the generalizations above, the percentage value (on the right of the fuzzy labels) tells to which extent it is possible to rely on the generalized tag. This helps to differentiate the precisions of hotness for different regions. Moreover, in the generalizations above it is clarified to which regions the word "regions" refers.

These generalizations can be used for dynamic drill-down and roll-up along any dimension to explore the desired patterns. The roll-up operation would correspond to the spatial generalization with precision value (fuzzy

TABLE 6. Generalizations with respect to dimensions

Generalizations with respect to temperature (and also season)
Cool (98%) regions (r6) that have the season winter (88%) cover small (99%) lands
Hot (85%) regions (r3, r4, r5) that have the season summer (85%) cover medium-sized (99%) lands
Mild (99%) regions (r1, r2) that have the season summer (83%) cover medium-sized (94%) lands
Generalizations with respect to season
Regions (r1, r2, r3, r4, r5) that have the season summer (85%) cover large (99%) lands
Regions (r6) that have the season winter (88%) cover small (99%) lands
Generalizations with respect to temperature (and also season) and precipitation
Hot (64%) regions (r5) with season summer (64%) and dry (100%) precipitation cover small (965) lands
Hot (99%) regions (r3) with season summer (99%) and fair (89%) precipitation cover small (97%) lands
Cool (98%) regions (r6) with season winter (88%) and wet (100%) precipitation cover small (99%) lands
Hot (99%) regions (r4) with season summer (99%) and wet (88%) precipitation cover small (98%) lands
Mild (98%) regions (r1, r2) with season summer (84%) and wet (97%) precipitation cover medium-sized (94%) lands
Generalizations with respect to season and precipitation
Regions (r5) that have the season summer (64%) and have dry (100%) precipitation cover small (96%) lands
Regions (r3) that have the season summer (%99) and have fair (89%) precipitation cover small (97%) lands
Regions (r1, r2, r4) that have the season summer (89%) and have wet (95%) precipitation cover medium-sized (98%) lands
Regions (r6) that have the season winter (88%) and have wet (100%) precipitation cover small (99%) lands
Generalizations with respect to precipitation
Dry (100%) regions (r5) cover small (96%) lands
Fair (89%) regions (r3) cover small (97%) lands
Wet (96%) regions (r1, r2, r4, r6) cover medium-sized (99%) lands

membership value) and the drill-down operation would correspond to the spatial specialization with precision values in the fuzzy spatial data cube. These operations can be performed easily on the generated generalized fuzzy spatial data, to see the nonspatial attributes and their precisions of the generalized spatial regions and details of the subregions.

4. Fuzzy Association Rule Generation from Fuzzy Spatial Data Cube

All generalization rules may not always be interesting. In order to obtain the interesting knowledge from the fuzzy spatial data cube it would be wise to generate fuzzy association rules from them.

Generating association rules from the fuzzy spatial data cube would be very useful, since the spatial data have computed precision values for their fuzzy dimensions and measures and fuzzy association rule mining is more easily computed than the spatial association rule mining. In fuzzy data cubes, the interesting association rules can be determined according to their significance and certainty factors, which reflect the reliability to generalization, instead of support and confidence factors which reflect the frequency of the data. In fuzzy spatial data cubes, it is also assumed that the reliability to generalizations is more important than the frequency of the data. Hence, this will help not to miss the rules which are not frequent but significant.

In the fuzzy spatial data cube constructed in the previous chapter, tuples in which the interested item-set occurs are already aggregated to a high-level tuple. Membership values of the dimensions of that tuple are more realistic since they are computed by taking the weighed measure value in account. Here, the term "item-set" corresponds to values of the "group by expression," i.e., "hot, dry" for "group by temperature, precipitation." In the fuzzy spatial data cube, knowledge that will be obtained by these group-by expressions is interested and valuable.

Considering the fuzzy association rule tried to be generated from the data in Table 2 where X = {temperature, precipitation}, A = {hot, dry}, Y = {area}, B = {large}, the significance was computed in (Kuok, 1998) as $S_{(X,A)} = (0.9 \times 0.2 + 0.7 \times 0.4 + 0.8 \times 0.3)/3 = 0.23$. In the fuzzy spatial data cube, there would already be an aggregated record for "hot, dry" item-set which was going to be obtained by the "group by temperature, precipitation" expression. The membership values of "hot" and "dry" fuzzy labels would be computed by taking into account the weighted measure value in the area. Hence, the vote of only that aggregated record would correspond to the significance of the item-set/group-by expression which is the multiplication of the membership values of the dimensions that take part in the group by clause.

Moreover, the certainty factor of the rule was defined in (Kuok, 1998) as $C_{((X,A),(Y,B))} = S_{(Z,C)}/S_{(X,A)})) = [(0.9 \times 0.2 \times 0.5 + 0.7 \times 0.4 \times 0.8 + 0.8 \times 0.3 \times 0.2)/3]/0.23 = 0.517$. Here, the significance of the antecedent and consequent is divided to the significance of the antecedent. In fuzzy spatial data cube, dimensions and measures are disjoint sets as the item-sets X and Y are defined in (Kuok, 1998). The significance of the antecedent and consequent for an association rule in the fuzzy spatial data cube would correspond to the multiplication of the membership values of dimensions and measures that take part in the rule respectively. Since the significance of rule is the multiplication of the membership values of the dimensions, the certainty of the rule would correspond to the multiplication of the membership values of measures, i.e., $(\Pi(\mu(ai)) \times \Pi(\mu(bi))/1)/(\Pi(\mu(ai)/1)$ for the aggregated tuple is $\Pi(\mu(b_i)$.

The significance and certainty factors for a fuzzy association rule that would be generated from the aggregated data in the fuzzy spatial data cube are:

$$\text{Significance} = \Pi(\mu(a_i)) \text{ where } a_i \subset A,\ 0 < i \leq |A|,\ X \text{ is a dimension},$$

$$\text{Certainty} = \Pi(\mu(b_i)) \text{ where } b_i \subset B,\ 0 < i \leq |B|,\ Y \text{ is a measure}.$$

The fuzzy association rules generated from the fuzzy spatial data cube with the threshold significance 0.85 and threshold certainty 0.9 are depicted in Table 7. Increasing the threshold values for significance and certainty factors would decrease the number of the generated fuzzy association rules.

TABLE 7. Fuzzy association rules

Antecedant => Consequent [Significance, Certainty]
Cool (0.97) => small (0.99) [0.97,0.99]
Hot (0.85) => mid (0.99) [0.85,0.99]
Mild (0.98) => mid (0.94) [0.98,0.94]
Winter (0.87) => small (0.99) [0.87,0.99]
Hot (0.99), fair (0.89) => small (0.97) [0.88,0.97]
Cool (0.97), wet (1.0) => small (0.99) [0.97,0.99]
Hot (0.99), wet (0.88) => small (0.98) [0.88,0.98]
Mild (0.98), wet (0.97) => mid (0.94) [0.96,0.94]
Summer (0.99), fair (0.89) => small (0.97)
Winter (0.87), wet (1.0) => small (0.99) [0.87,0.99]
Dry (1.0) => small (0.96) [1.0,0.96]
Fair (0.89) => small (0.97) [0.89,0.97]
Wet (0.95) => mid (0.99) [0.95,0.99]

The association rule "hot (0.99), wet (0.88) => small (0.98) [0.88,0.98]" can be commented as "Given that an area is hot (0.99) and has wet (0.88) precipitation, it can be concluded that it covers small piece of land with the certainty %98 while the significance of that area being both hot and wet is %88."

Furthermore, comparing fuzzy association rules obtained at different times would be helpful in determining the deviations seen along time. For example, considering the fuzzy association rules "In 2003, hot (0.99), wet (0.84) => small (0.98) [0.83, 0.98]" and "In 2004, hot (0.83), wet (0.92) => small (0.7) [0.76, 0.7]"; it can be concluded that the degree of hotness of the temperature has decreased in 2004, while the degree of wetness of the precipitation has increased and hot and wet areas have enlarged. It can also be concluded that the probability of regions being both hot and wet has decreased and given that a region is hot and wet it is less probable that it covers small area.

5. Case Study: "Weather Pattern Searching"

A case study has been implemented in order to illustrate how a fuzzy spatial data cube can be constructed and how fuzzy association rules can be generated from it for weather pattern searching.

The fuzzy spatial data cube construction was implemented in Java as a Java Applet, in Eclipse 2.1 environment with JRE 1.4.2_04 and on Microsoft SQL Server 2000 database. These two environments were combined by

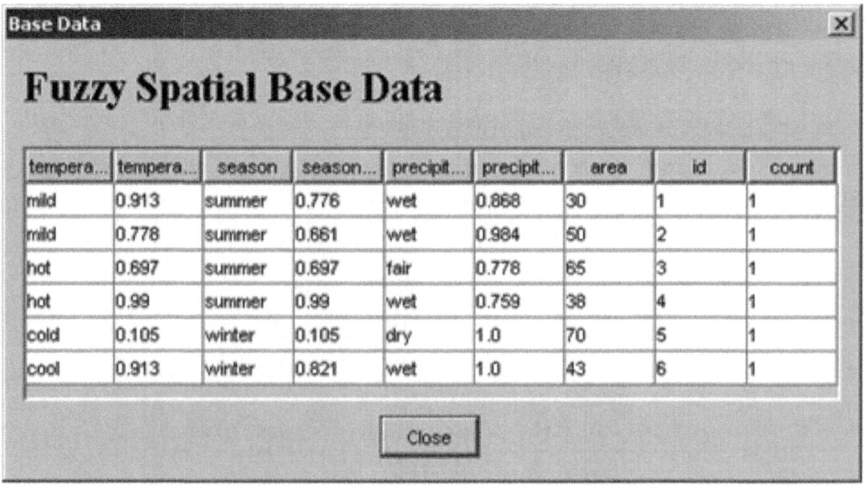

tempera...	tempera...	season	season...	precipit...	precipit...	area	id	count
mild	0.913	summer	0.776	wet	0.868	30	1	1
mild	0.778	summer	0.661	wet	0.984	50	2	1
hot	0.697	summer	0.697	fair	0.778	65	3	1
hot	0.99	summer	0.99	wet	0.759	38	4	1
cold	0.105	winter	0.105	dry	1.0	70	5	1
cool	0.913	winter	0.821	wet	1.0	43	6	1

Figure 8. View of fuzzy spatial base data

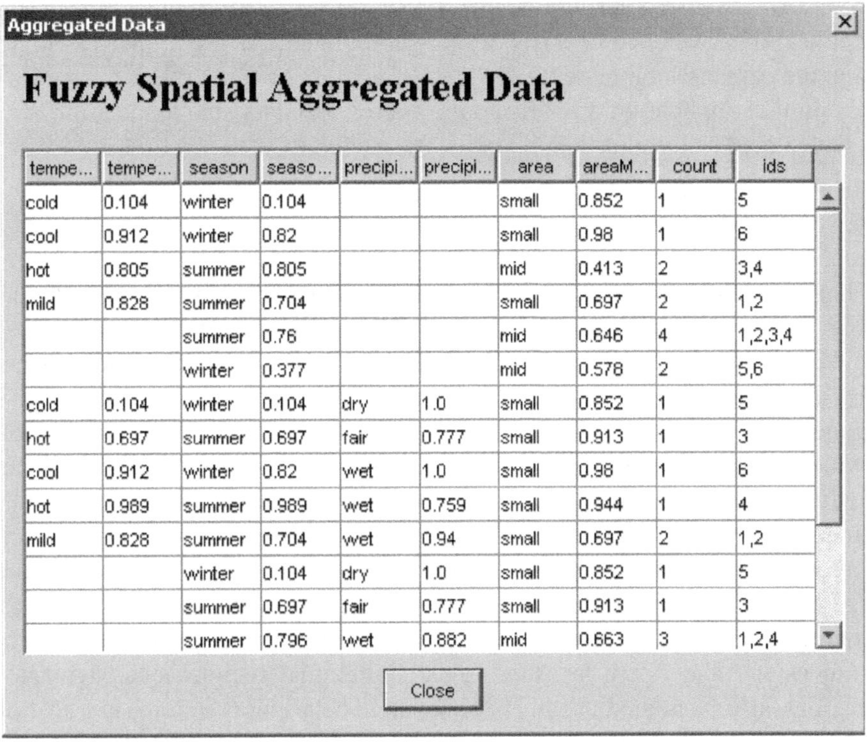

tempe...	tempe...	season	seaso...	precipi...	precipi...	area	areaM...	count	ids
cold	0.104	winter	0.104			small	0.852	1	5
cool	0.912	winter	0.82			small	0.98	1	6
hot	0.805	summer	0.805			mid	0.413	2	3,4
mild	0.828	summer	0.704			small	0.697	2	1,2
		summer	0.76			mid	0.646	4	1,2,3,4
		winter	0.377			mid	0.578	2	5,6
cold	0.104	winter	0.104	dry	1.0	small	0.852	1	5
hot	0.697	summer	0.697	fair	0.777	small	0.913	1	3
cool	0.912	winter	0.82	wet	1.0	small	0.98	1	6
hot	0.989	summer	0.989	wet	0.759	small	0.944	1	4
mild	0.828	summer	0.704	wet	0.94	small	0.697	2	1,2
		winter	0.104	dry	1.0	small	0.852	1	5
		summer	0.697	fair	0.777	small	0.913	1	3
		summer	0.796	wet	0.882	mid	0.663	3	1,2,4

Figure 9. View of fuzzy spatial aggregated data

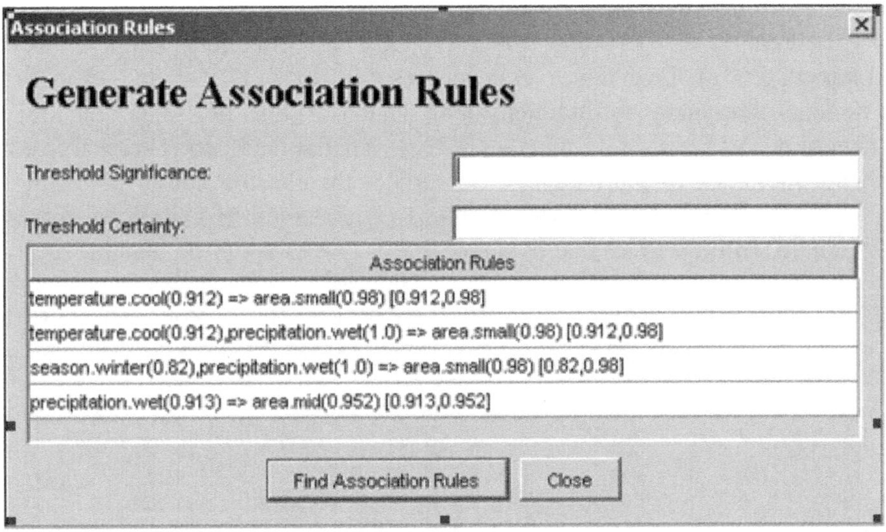

Figure 10. Display of fuzzy association rules

Microsoft SQL Server Driver for JDBC access. The architecture of the application developed consists of three components as GUI, business logic, and the database component.

In that application, fuzzified data, aggregated data, and association rules are depicted as in Figures 8, 9, and 10 respectively.

6. Discussion

Han has done some studies for spatial generalizations included in [6] and Stefanovic et al. have extended these studies by constructing spatial data cubes included in (Stefanovic, 2000). But the generalization in their studies is imprecise as shown in Figure 11 below. For example, they have generalizations like "cold and dry regions (A04, T90 ...) are 200,000 m^2 wide." Here, the preciseness of hotness and dryness are missing both for individual regions and for the aggregated region. Moreover, in Han's and Stefanovic's study, the measures are neither generalized nor their preciseness is computed and indicated. In this thesis, that is also considered.

This study mainly focuses on how preciseness of the generalized values can be considered as illustrated in Figure 12. Membership values for fuzzy dimensions and fuzzy measures of all individual regions and aggregated regions are calculated in the fuzzy spatial data cube. Before aggregation is done, spatial data are generalized and precision for generalization is calculated. Then, individual regions with common generalized values (i.e., hot) are aggregated together. Additional calculation is done for the precision of generalized aggregated spatial data.

Laurent et al. (Laurent, 2000; Laurent and Bouchon-Meunier, 2001; Laurent, 2001) have done some studies for fuzzy OLAP and proposed a model for fuzzy multidimensional databases that are used for fuzzy summaries generation. In this thesis, fuzzy spatial data cube is constructed; hence the usage of spatial data differentiates the construction of such cubes from Laurent's study at some points. In their study, a value can belong gradually to more than one concept, that is, Ankara can be generalized to

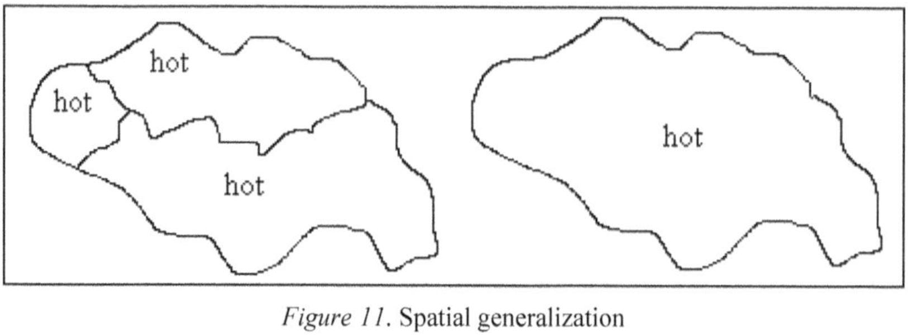

Figure 11. Spatial generalization

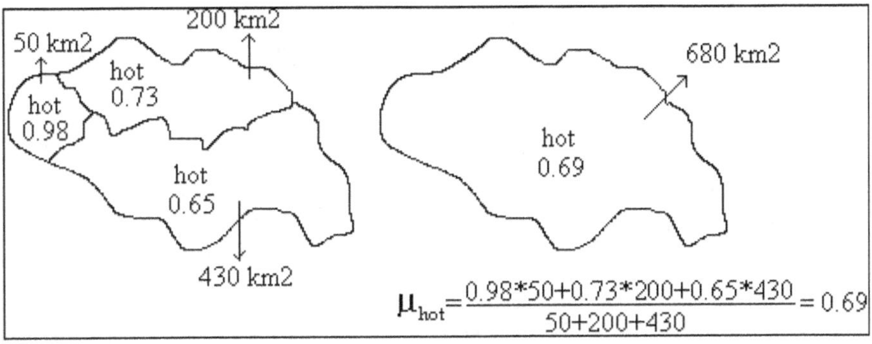

Figure 12. Fuzzy spatial generalization

take part in the east of Turkey with the precision 0.4 and it can also be generalized to take part in the west of Turkey with the precision 0.5. On the other hand, in the fuzzy spatial data cube, a value can be generalized to only one fuzzy label, to the label to which it belongs with the highest membership. Membership values for all defined fuzzy sets are computed and the maximum membership value and its corresponding fuzzy set are considered. That is important from the point of hierarchies. It is assumed that a value can be generalized to only one upper level since this makes fuzzy spatial generalization rules more sensible. In Figure 12, it is supposed that the three subregions on the left, fit best to the "hot" fuzzy label and they are aggregated to the larger region on the right.

In Laurent's study, each slice corresponds to the cube with a membership value, that is, a value of one dimension has the same membership value for all the cells in the slice. In other words, in the fuzzy data cube, aggregated cells along one slice and individual cells along that slice have the same membership value for one-dimensional value. For example, the slice corresponding to sales done for oven along the production dimension can belong to the fuzzy data cube with degree 0.7. But in fuzzy spatial data cube each cell has its individual membership value for the corresponding dimension value since spatial objects might have common properties but each spatial object might have that property with a different degree than other spatial objects. Membership values of all cells along a dimension value are considered during aggregation in fuzzy spatial data cube. Fuzzy spatial data cube is semantically different than fuzzy data cube.

Furthermore, in Laurent's study, arithmetic average of membership values of cells is computed when fuzzy data are summarized, i.e., $\mu hot = (0.98 + 0.73 + 0.65)/3 = 0.79$. But regarding spatial data, it is not reliable to only take the arithmetic average since attributes of the spatial objects depend on space (or other issues as population). The way of computing the membership value of the aggregated region in fuzzy spatial data cube is more realistic, since it will have a value closer to the value of the largest region (or most crowded). Taking arithmetic average may cause to some wrong deviation as it can be easily concluded by comparing results.

Fuzzy spatial data cubes, make it possible to track deviations in precisions of characteristic properties of spatial data which was not an issue neither in spatial data cubes nor in fuzzy data cubes. For example, comparing generalization rules such as "Ankara was %80 hot and %78 dry in June, 2003" and "Ankara was %85 hot and %96 dry in June, 2004" enables seeing deviations in precisions of hotness (5% increase) and dryness (18% increase) values for Ankara between June 2003 and June 2004. Changes in spatial characteristics are easily identified in fuzzy spatial data cubes and this helps in decision-making about spatial data.

Another important issue is that, fuzzy association rules are generated from the fuzzy spatial data cube since the spatial data have precision values for the fuzzy dimensions and measures. Computations of significance and certainty factors are modified according to the aggregated data in the fuzzy spatial data cube. In previous studies, a fuzzy association rule is mined over many tuples in a relational database, but in fuzzy spatial data cube it can be mined over one aggregated tuple. Hence, fuzzy association rule mining is more feasible in fuzzy spatial data cube than it is in relational databases. Additionally, when the reliability to generalizations is more important than the frequency of the data in order not to miss the infrequent but significant rules, fuzzy association rule mining is more easily computed over spatial data than the spatial association rule mining, which has higher computational complexity.

7. Conclusion

In this study, the concepts of fuzzy data cubes and spatial data cubes are combined to get benefit from both of them. A new method is proposed for the aggregation of the fuzzy dimensions and measures and their memberships regarding spatial data. Moreover, in this study, aggregation on fuzzy hierarchies is handled, which helps in obtaining higher levels of generalizations.

The constructed fuzzy spatial data cube can be used for generating fuzzy association rules. The way the aggregation is handled for fuzzy spatial cube in this study enhances the generation of fuzzy association rules from transactional databases. Computation of the significance and certainty factors is done more easily and more realisticly since the memberships of the dimensions for the aggregated spatial objects are computed regarding the weighted measure.

To sum up, this study contributes to the computational area by introducing fuzzy logic to spatial data cubes and proposing a new method for the aggregation of the fuzzy spatial attributes. The obtained generalization rules help in commenting the core spatial data and concluding decisions about it. Moreover, it is shown that fuzzy spatial data cubes can enhance the generation of fuzzy association rules from spatial data which increases the quantity and quality of the knowledge about the spatial data.

Future studies can be done about how fuzzy logic can be used for enhancing prediction of unknown and missing spatial data in spatial data cubes, or for comparing different classes of data and perform relevance analysis to find attributes that best distinguishes different classes, or classify spatial data and construct a model and use it to classify new data, or for clustering spatial data to find distribution patterns.

References

S. Chaudhuri, U. Dayal, An Overview of Data Warehousing and OLAP Technology ACM SIGMOD Record, Vol. 26, pp. 65–74 (1997).

M. Stonebraker, J. M. Hellerstein, Readings in Database Systems, Morgan Kaufmann Publishers (1998).

M. T. Özsu, P. Valduriez, Principles of Distributed Database Systems, Prentice Hall (1999).

J. Han, M. Kamber, Data Mining: Concepts and Techniques, Academic Press (2001).

R. Jacobson, Microsoft SQL Server 2000 Analysis Services Step by Step, Microsoft Press, (2000).

W. Lui, J. Han, Discovery of general knowledge in large spatial databases, Proceedings of 1993, Far East Workshop on GIS, pp. 275–289 (1993).

P. Rigaux, M. O. Scholl, A. Voisard, Spatial Databases with Application to GIS, Morgan Kaufmann Publishers Incorporation (2002).

O. R. Zaiane, Multimedia and spatial data mining, Principles of Knowledge Discovery in Data, University of Alberta (2002).

N. Stefanovic, J. Han, K. Koperski, Object-based selective materialization for efficient implementation of spatial data cubes, IEEE Transactions on Knowledge and Data Engineering, Vol. 12, No. 6 (2000).

V. Harinarayan, A. Rajaraman, J. D. Ullman, Implementing data cubes efficiently, Proceedings 1996, ACM-SIGMOD, International Conference on Management of Data, pp. 205–216 (1996).

J. Gray, A. Bosworth, A. Layman, H. Pirahesh, Data Cube: A Relational Operator Generalizing Group-by, Cross-tab, and Roll-up., Proceedings of the 12th International Conference on Data Engineering, pp. 152–159 (1996).

S. Agarwal, R. Agrawal, P. M. Deshpande, On the computation of multidimensional aggregates, Proceedings of the 22nd VLDB Conference (1996).

Y. Zhao, P. Deshpande, J. F. Naughton, An Array-based algorithm for simultaneous multidimensional aggregates, Proceedings ACM SIGMOD International Conference on Management of Data, pp. 159–170 (1997).

A. Laurent, B. Bouchon-Meunier, A. Doucet, Fuzzy data mining from multidimensional databases, International Symposium on Computational Intelligence, Studies in Fuzziness and Soft Computing, 54: 278-283 (2000).

A. Laurent, Generating fuzzy summaries from multidimensional databases, Fourth International Symposium on Intelligent Data Analysis, pp. 24–33 (2001).

A. Laurent, B. Bouchon-Meunier, A. Doucet, Towards fuzzy-OLAP mining, Proceeding Work. PKDD Database Support for KDD, pp. 51–62 (2001).

J. Han, OLAP mining: an integration of OLAP with data mining, Proceedings of the 7th IFIP2.6 Working Conference on Database Semantics (DS-7), pp. 1–9 (1997).

B. Kelkar Exploiting symbiosis between data mining and OLAP for business insights, DM Direct Newsletter (2001).

C. White, Intelligent business strategies: OLAP in the database, DM Review Magazine (2003).

Pilot Software Acquisition Corporation, An Introduction to OLAP Multidimensional Terminology and Technolog, http://www.pilotsoftware.com/pdf/olapwp.pdf (2002).

X. Zhou, S. Prasher, M. Kitsuregawa, Database support for spatial generalization for www and mobile applications, Third International Conference on Web Information Systems Engineering, Singapore (2002).

S. Prasher, X. Zhou, Multiresolution amalgamation: dynamic spatial data cube generation, In Proceedings of Fifteenth Australian Database Conference (ADC 2004), pp. 103–111 (2004).

N. Stefanovic, Design and Implementation of On-Line Analytical Processing (OLAP) of Spatial Data, master's thesis, Simon Fraser University, Canada (1997).

J. Han, K. Koperski, N. Stefanovic, GeoMiner: a system prototype for spatial data mining, Proceedings of 1997 ACM-SIGMOD, International Conference on Management of Data, pp. 553–556 (1997).

D. Papadias, P. Kalnis, J. Zhang, Y. Tao, Efficient OLAP operations in spatial data warehouses, Seventh International Symposium on Spatial and Temporal Databases (2001).

X. Zhou, D. Truffet, J. Han, Efficient polygon amalgamation methods for spatial OLAP and spatial data mining, Proceedings of the Sixth International Symposium on Large Spatial Databases (1999).

Y. Bedard, T. Merret, J. Han, Fundamentals of spatial data warehousing for geographic knowledge discovery, Geographic Data Mining and Knowledge Discovery, CRC Press (2001).

F. Petry, Introduction to Fuzzy Databases Kluwer Publisher (1996).

C. M. Kuok, A. Fu, M. H. Wong, "Mining Fuzzy Association Rules in Databases", ACM Sigmod Record, Vol. 27, pp. 41–46 (1998).

INTERACTIVE OBJECTS EXTRACTION FROM REMOTE SENSING IMAGES

VICTOR BUCHA*
United Institute of Informatics Problems of National Academy of Sciences of Belarus, Surganova 6, 220012 Minsk, Belarus; bucha@newman.bas-net.by

SERGEY ABLAMEYKO
United Institute of Informatics Problems of National Academy of Sciences of Belarus, Surganova 6, 220012 Minsk, Belarus; abl@newman.bas-net.by

Abstract. A high-level interactive approaches for objects extraction (roads, rivers, forests, buildings, etc.) from remote sensing images and color-scanned maps are proposed based on connected components accumulation and live-wire techniques. In contrast to known interactive segmentation techniques, the proposed approach allows extraction of object centerline as well as boundary. The obtained experimental results are very promising and show that the proposed approach allows us solving object extraction task with high speed and quality.

Keyword: interactive segmentation, cartographic objects extraction, live-wire techniques, remote sensing images interpretation, vectorization, pixel force field

1. Introduction

Many researchers concentrate their efforts on automatic objects extraction with subsequent hand correction of results (Laptev et al., 2000; Mena, 2003). In fact, fully automatic approaches often provide the GIS with unsatisfactory imperfect data and correction time can be comparable with semiautomatic digitization. Therefore interactive recognition approaches are developed to take upon oneself the routine work of operator while keeping the possibility of full user control of segmentation/recognition process.

*To whom correspondence should be addressed. Victor Bucha, United Institute of Informatics Problems of National Academy of Sciences of Belarus, Surganova 6, 220012 Minsk, Belarus. e-mail: bucha@newman.bas-net.by

A. Morris and S. Kokhan (eds.), Geographic Uncertainty in Environmental Security, 225–238.

Interactive segmentation techniques can be either region-based or boundary-based (Falco and Udapa, 1998). The magic wand tool (Dayton and Davis, 2001) belongs to the first group and enables user to interactively select a seed point to grow a region by adding adjacent neighboring homogeneous pixels. Active contour (also called snake) is a well-known boundary-based technique which allows setting an initial curve to be modified by the external and internal forces (Laptev et al., 2000). The final boundary is estimated while minimizing energy functional.

Another well-known boundary-based technique is live wire (Falco and Udapa, 1998) which uses a global graph search. The pixels are considered like graph nodes, and a cost based on boundary features is assigned to graph arcs. A minimum cost path from the seed to every image pixel is calculated. User gets the desired segmentation result by interactively moving a cursor near object's boundary.

Such approaches, however, can extract only the area objects while GIS needs a center line for elongated objects. In addition, it is very difficult for the magic wand and snake approaches to provide user with the interactive visual feedback.

In this paper, we consider two approaches for interactive objects extraction: one based on connected components accumulation and one based on live-wire paradigm.

In contrast to known interactive segmentation techniques, the proposed approach based on live-wire paradigm allows extraction of object's centerline rather than boundary.

The centerlines are extracted using a novel image representation scheme called pixel force field (PFF). PFF transformation is considered as a generalization of a distance transformation (DT) (Jang and Hong, 2001) for color and grayscale images using not only distance but also qualitative pixel features such as color and gray scale. The skeleton features representing a centerline of elongated objects (roads, rivers, etc.) are highlighted by PFF transformation and used by live-wire technique for objects extraction from remote sensing images.

2. Desiderata

Remote sensing images conversion (interpretation) to digital maps is a demanding task; a very small number of errors can have a large effect on the usability of an interpretation. It is now widely accepted that completely automatic, error-free interpretation of large and complex remote sensing images and maps will not be achieved in the short to medium term and that user involvement is likely to remain a part of interpretation for some time to come.

Fully manual digitization is however, also unacceptable. It places very high demands on operators, leading to a large number of input errors. A further drawback of traditional manual methods is the difficulty of providing visual control to the digitization process. The user typically finds it very hard to tell which objects have been digitized and which are awaiting input.

The need for integrated manual and automatic drawing interpretation is a recurring topic in workshop discussions and in the conclusions of published papers (Bucha et al., 2005; Pridmore et al., 2005). Most recently developed systems include the user, but are either primarily automatic systems allowing only limited user involvement, usually in the form of post hoc editing, or are essentially manual.

Pridmore et al. have recently based analyses of line drawing interpretation (Darwish et al., 2001) and aerial image understanding [Pridmore et al., 2005] systems on the inference structures associated with KADS/CommonKADS. The following statements can be made on the basis of his research (Pridmore et al., 2005; Darwish et al., 2001).

The provision of editors with which the results of interpretation can be manually improved is commonplace. Truly interactive interpretation systems should, however, give the user the feeling that he/she is collaborating in the process, not just correcting errors made by a poor quality system. Users' inputs should therefore be simple in form and seen to have significant effect.

Simplicity emphasizes that high-level labels can be applied to and relations stated or verified between drawing constructs using fewer user input operations than are required, for example, to edit spurious blobs and holes from a binary image. Having significant effect implies that the user should intervene when the information s/he can provide would affect the interpretation of an extended area of the drawing or the identification of a physically large or semantically important construct. While low-level decisions can have important consequences, this again suggests that much of the user's interaction should involve the higher levels of the system. Where lower level interaction is needed the requirement to provide feedback makes it particularly important that users have a clear understanding of why their input is necessary.

Where possible systems should avoid asking for input unless it is both necessary and the reason for it can be conveyed to the user via the interface. The user should however, have the ability to be proactive. When an automatic operation is seen to introduce errors and/or distortions, the user must be able to interrupt the system and correct the error while retaining the results of any valid processing performed.

The designer of an interactive interpretation system must decide where in the process the user will have opportunities to interact with the system and what guidance and feedback should be embedded in the interface. To do this requires detailed and explicit knowledge of the information available at each point in the process and the decisions likely to be affected by the introduction of new information.

3. Objects Extraction Approach Based on Connected Components

In most of the automatic systems (Ablameyko and Pridmore, 2000), the process of scanned map digitizing can be divided into three main stages: global binarization, raster-to-vector transformation with the aim of obtaining a structural representation, and automatic/interactive recognition of cartographic objects to obtain the required final map representation.

This way is not always optimal because after global binarization some information is already lost and can not be reconstructed and used. Another problem that user should return and repeat previous stage if the results were not satisfactory. In addition, it is extremely awkward to work with binary image rather than color one, since some entity connections are not evident on binary image and user should guess which leads to errors.

Therefore an approach for object extraction based on connected components accumulation is proposed which involves user more deeply in low-level interpretation and allows direct interpretation on original color image without noticeable increasing of time costs.

The proposed system (Figure 1) comprises five stages: local color separation, area and line object detection, local thinning or contour extraction, object tracing, and automatic object vectorization. In addition, tools for seed point snapping and real-time control of the process are available.

First, an initial segmentation is performed. The operator specifies a seed point and color tolerance parameters, which are used to define a boundary condition of the region to be sought. The region-based algorithm described in (Smith, 1979) is applied, growing a connected component of near-uniform color by adding suitable neighboring pixels to the seed. This algorithm is both memory- and computationally efficient.

Manually placing a seed point inside the required region is tedious and often difficult, particularly when small components are to be extracted. If the chosen seed lies outside the object or near an object boundary frustrating errors can occur. To facilitate seed point placement, cursor snap facility and real-time region highlighting are provided. The snapping tool highlights and/or forces the mouse cursor to the nearest minimum gradient

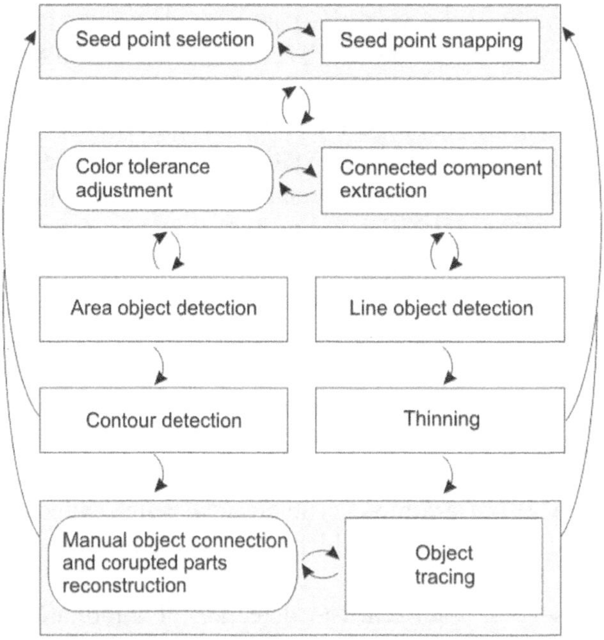

Figure 1. The architecture of the interpretation system based on connected components

magnitude pixel within a user specified neighborhood. Thus, as the mouse cursor is moved by the user, it snaps to nearby, potentially "good" seed pixels. If the cursor stops at any pixel for a time T, then the connected component which would be extracted using that pixel as a seed and the current parameter set is highlighted.

Once a connected component has been identified, the class (line or area) of that component is determined automatically. This stage is necessary since area and line objects are represented in different ways and separate algorithms are required to form each representation. A bounding contour describes area objects while skeletons are extracted from line entities.

Either local thinning or contour detection is then performed, depending on the class of the object in question. The local nature of this processing distinguishes this step from the globally applied algorithms used in fully automatic digitization systems. Focusing on a connected component selected by the user reduces the memory and computational resources required, making real-time display and control of the process practical. The thinning algorithm described in (Lei et al., 2003) is employed here.

The next stage is object tracing, in which distinct components are semiautomatically combined to produce descriptions of objects. Given two line components, the system first attempts to find a path, lying entirely within them, which connects their seed points. If such a path is found, the two regions are combined and vectorized as a single entity. If not, user must decide how to connect the regions. There are two possibilities. He/she may enter further seed points, producing an additional connected components linking the existing ones. Alternatively, the corrupt area may be reconstructed in manual mode. Classical vectorization and approximation is used to produce the desirable vector representation. In addition, information (color, thickness), about the underlying connected component(s) can be recorded. Area objects are combined using object-oriented mathematical morphology.

3.1. LINE AND AREA OBJECT DETECTION

Without the ability to distinguish different object types there is a danger that important information will be lost when an inappropriate representation is computed. Skeletons do not provide an adequate representation of area objects (Figure 2) but are more efficient and better suited to line objects than contour-based techniques, which allow fast editing and coding of area objects.

There are several approaches to detection of information loss through inappropriate representation and the classification of area and line objects (Lei et al., 2003; Ablameyko and Pridmore, 2000). The technique reported in [Lei et al., 2003] detects information loss during a thinning procedure by applying a threshold to the value R

$$R = \frac{Area[skeleton]}{Area[contour]},$$

where Area[] is an operator that counts the number of pixels in a component. If R is lower than the threshold, a contour representation is chosen over a skeleton. However to estimate the threshold value the contour and skeleton representations must both be estimated. This is too time consuming to employ within a semiautomatic system where real-time feedback is required.

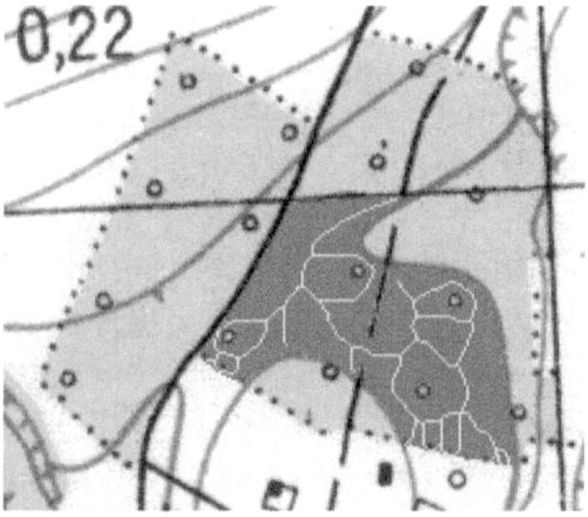

Figure 2. Thinning area objects

Instead, we based area and line object recognition on the following measure, which does not require thinning to be performed (Ablameyko and Pridmore, 2000).

$$w = 0.25 \cdot (P - \sqrt{P^2 - 16 \cdot S})$$

The perimeter P and square S of a connected component can be efficiently obtained on the fly during initial segmentation. Depending on the value of w, a contour or line representation is extracted.

3.2. TRACING LINE OBJECTS

Consider the situation in which two line objects overlap (Figure 3A). Suppose the operator's goal is a vector representation of a line linking pixels E1 and E2 (Figure 3A). Two seed pixels, S1 and S2, are specified (Figure 3A), and two connected components are extracted (Figure 3B). If a connected path can be achieved between S1 and S2 then vectorization can be done automatically without further user intervention. Overlaps and intersections are however, common. Additional path reconstruction processes will often be required.

First, thinning applied and the skeleton of each component is estimated (marked as a thin white line in Figures 3 and 4). Next, feature pixels are found. We define the following types of feature pixel:

End pixels
Node pixels
Connected pixels

When feature pixels have been marked, the path between seed points can be found. We examine the neighborhood of each end point to find places where a line approximation should take place. For each end pixel

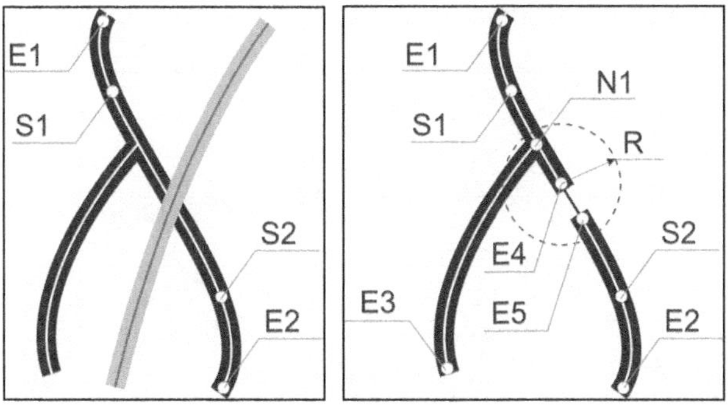

(A) (B)

Figure 3. Interaction of two objects (A) and extracted components (B)

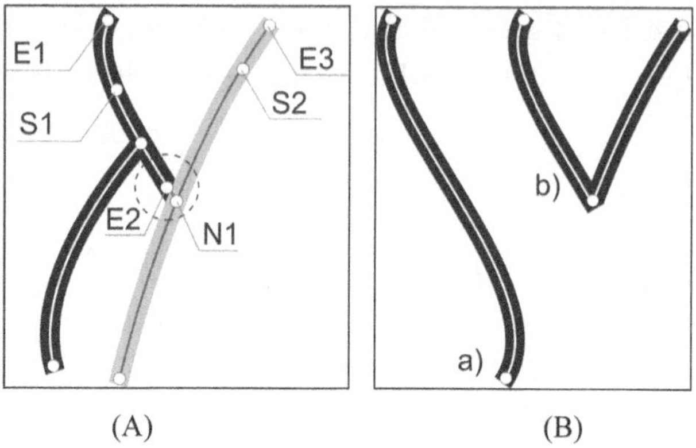

Figure 4. Tracing a line object (A) and result of digitization (B)

belonging to one connected component the distance to the nearest pixel on the skeleton of another component is computed. A line then connects the pair of pixels with shortest distance between them (Figure 3B). An upper limit R is placed on the length of these connecting lines. At any time the user can decide if and how to create a line between pixels: to select the most appropriate path, to vary R or to switch to manual approximation mode. The results of applying this process to the examples are shown in Figure 4B.

3.3. TRACING AREA OBJECTS

Given area objects, the goal is to combine regions. Suppose the operator wishes to digitize the green area of (Figure 5A). Two seed pixels produce two connected components (Figure 5B). Mathematical morphology can be used to combine these components (Figure 6A).

The classical operations of closing and opening however, are not suitable as they corrupt the boundary of the resulting object (Figure 6B). The black pixels in Figure 6B mark the affected pixels. To avoid this problem we use object-oriented morphological operators.

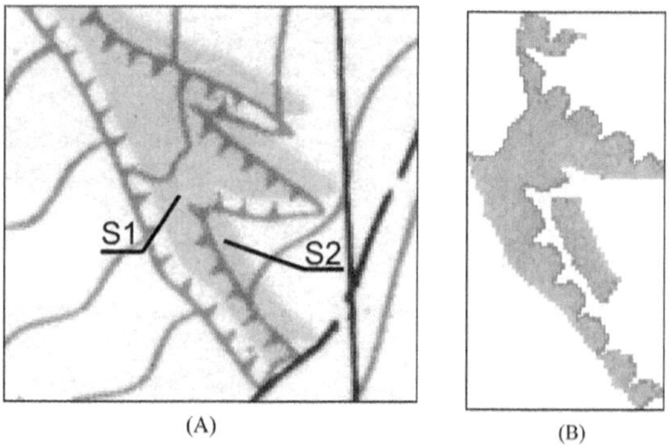

Figure 5. A map image with seed pixels (A) and the connected components extracted (B)

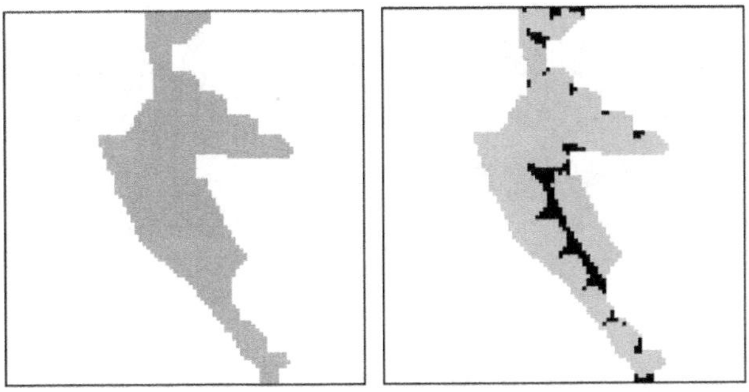

Figure 6. Combined object (A) and difference between original components (B)

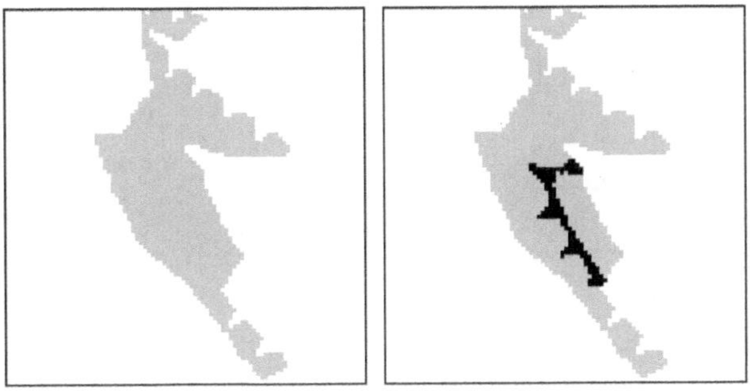

Figure 7. Result of object-oriented opening (A) and difference between original components (B)

Object-oriented morphological operators are executed only when pixels arising from different connected components are under consideration. In other words, the morphological operators do not affect pixels of the same component. This prevents the corruption of external and internal boundaries of components. The result of the operation is shown in Figure 7.

3.4. EXPERIMENTAL RESULTS

The system described here has been used to digitize area and line objects in a variety of remote sensing images (e.g., Figure 8). Figure 8B shows the boundaries of area and skeletons of line objects extracted in this way.

Comparisons of processing times and boundary accuracy achieved have been made between the proposed approach and a standard manual tracing method. Using the proposed approach the same extraction took at most half times needed for manual digitization. The accuracy of the boundary and skeletons extracted by the proposed approach was also better in each case. When properly targeted, automatic thinning, contouring, and entity extraction are more stable than manual tracing.

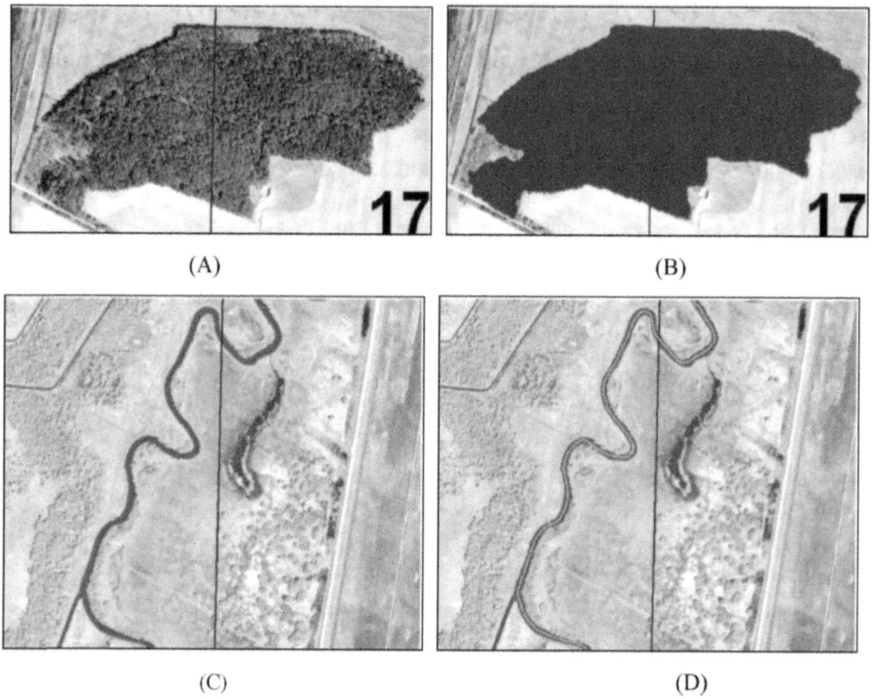

Figure 8. Area and elongated objects (A, C) and extracted boundary (B) and centerline (D)

4. Objects Extraction Approach Based on Live-Wire Paradigm

There are a number of interactive approaches of interactive objects extraction. Among those approaches, we inspired on segmentation paradigm live wire (Falco and Udapa, 1998) and develop a semiautomatic road extraction technique at remote sensing images based on pixel force field.

In the proposed method, a user first specifies a start point and an end point of the target road on a bitmap image. This is the only step manually performed by the user. The remaining steps will be done automatically. Then a regular-mesh graph is prepared on a bitmap image by treating every pixel as a node and putting an edge w_{pipj} between every pair of adjacent nodes (Figure 9).

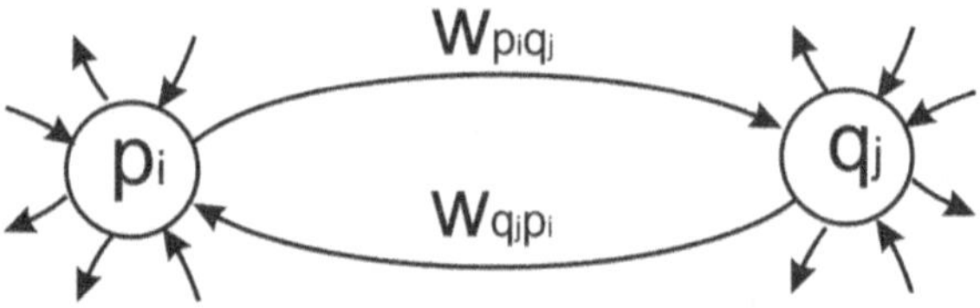

Figure 9. Graph model

Now, the elongated object extraction problem can be considered an optimal path problem between the start and the end nodes. Thus, for every edge of the graph, a weight is assigned and total weight accumulated along the path is to be minimized.

The weight is assigned with the proposed pixel force field (Ablameyko and Bucha, 2005). The simplest definition of the criterion is the magnitude of the force vector at each pixel, because the force vector F_i often becomes small around the skeleton. Thus, under this criterion, the elongated object extraction task is similar to the foregoing thinning task (i.e., skeletonization task), although road extraction task is organized in an optimization framework and provides connected path as a road. The graph weights w_{pipj} are assigned with following equation:

$$w_{p_i,p_j} = \left|F_{p_i}\right|,$$

Using a dynamic programming (DP)-based algorithm [Falco and Udapa, 1998], we can obtain the minimum weight path between the start point and the end point efficiently. It is noteworthy that the DP algorithm can provide the minimum weight paths between a fixed start pixel and any other pixels at once. In other words, by only specifying a start pixel manually, the user can obtain the minimum weight path to any end pixel. In this sense, we refer to the start pixel as "seed point" and refer to the end pixel as "free point". Figure 10 shows an example of the result of road extraction from a remote sensing image.

The optimization procedure discussed above can be utilized for boundary detection from color images. The modification from the elongated object extraction task is done only in the criterion. Specifically, if we evaluate "boundariness" at each edge of the mesh graph, we can obtain the boundary of the image under an optimization framework.

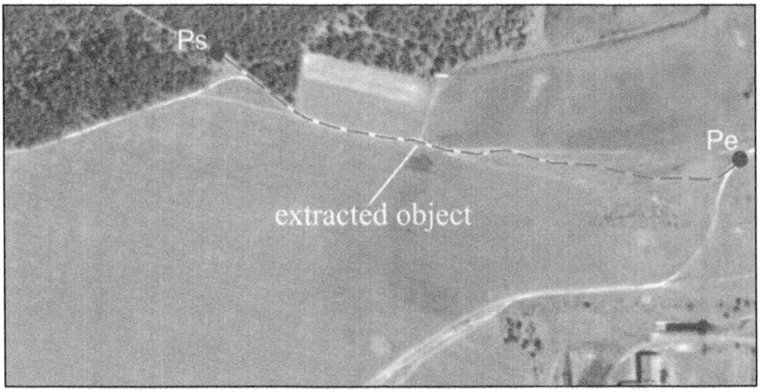

Figure 10. Result of road extraction (Ps: seed point. Pe: free point)

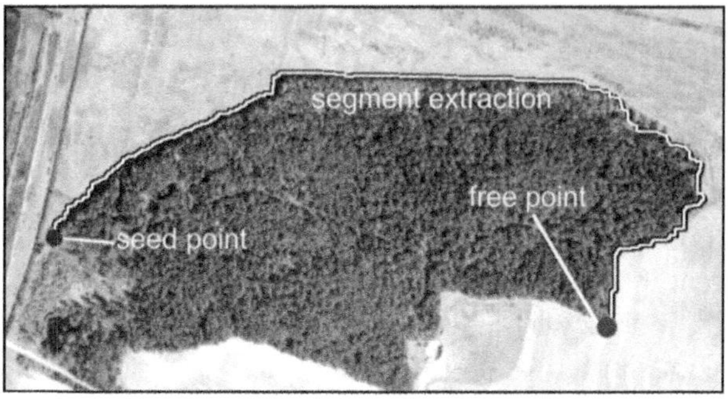

Figure 11. Example of boundary detection

Pixel force field can be used for the evaluation of boundariness. The boundariness of the pixel P_i can be evaluated by finding minimum weight paths between a fixed start pixel and any other pixels. The graph weights w_{pipj} are assigned with following equation:

$$w_{p_i, p_j} = -\left|F_{p_i}\right|,$$

Figure 11 shows an experimental result of boundary detection. After the seed and the free points were specified, the boundary between the forest area and the surrounding field area was detected accurately.

4.1. EXPERIMENTAL RESULTS

Figure 12A shows a result of road portion extraction from a remote sensing image. A road running across the street is extracted successfully with indication of four points. Hidden in shadow and distorted with noise roads are extracted as well (Figure 12B) whereas other semiautomatic and automatic approaches do not cope with.

(A) (B)

Figure 12. Result of road (A) and hidden road extraction (B)

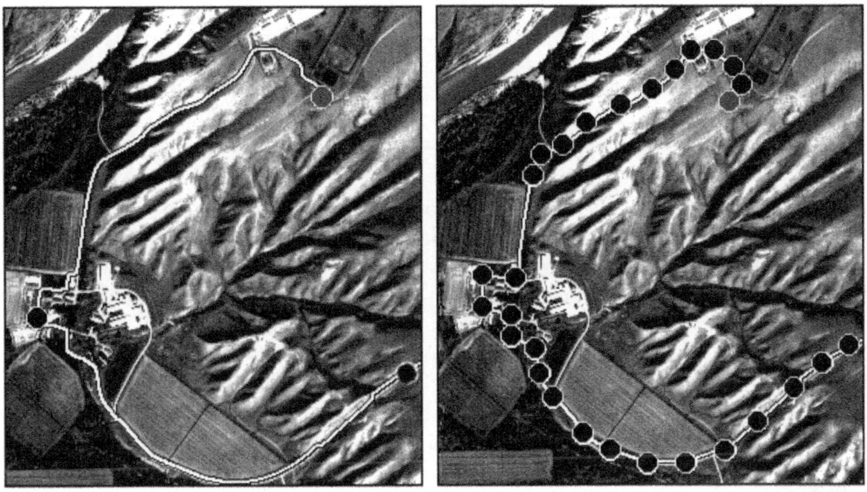

(A) (B)

Figure 13. Road extraction with live-wire 3 points, 6 sec (A) and hand mode 29 points, 66 sec (B)

In comparison with snake technique (1), which needs rough definition of extracted object and its post correction in a case of failure, the proposed approach allows extraction without post correction step.

Figures 13A and 13B compare a performance of proposed approach with hand extraction of roads. The number of indicated points and average time extraction is reduced in approximately 10 and 3.5 times accordingly.

In addition, the center line of elongated objects can be extracted in similar way for color, gray scale, and multispectral remote sensing images.

5. Conclusion

The proposed approaches provide accurate and efficient interactive tools for the extraction of objects from remote sensing images. The enhanced interactivity provided by approaches, supports high-quality digitization even in very difficult situations, when automatic systems fail.

The developed approaches are implemented into experimental software and allow to improve the level of automation and efficiency of digital map creation and update. As a result the cost for map and remote sensing images interpretation is also reduced.

References

Ablameyko, S., Bucha, V., 2005, Image pixel interaction and application to image processing, Pattern Recognition and Image Analysis, **15**(1): 136–138.

Ablameyko, S., Pridmore, T., 2000, Machine Interpretation of Line Drawings Images, Springer.

Bucha, V., Ablameyko, S., Pridmore, T., 2005, Intellectual semi-automated vectorization of multicolor cartographic objects, Visual information engineering, Proceedings of IEE International Conference, University of Glasgow, pp. 115–120.

Darwish, A., Pridmore , T., Elliman, D., 2001, Interpreting aerial images: a knowledge-level perspective, Proceedings ES-2001, Cambridge, pp. 169–182.

Dayton, L., Davis, J., 2001, The Photoshop 6 Wow!, Peachpit press.

Falco, X.A., Udapa, J.K., 1998, User-steered image segmentation paradigms: live wire and live lane, Graphical Models and Image Processing, **60**: 233–260.

Jang, J.-H., Hong K.-S., 2001, Linear band detection based on the Euclidean distance transform and a new line segment extraction method, Pattern Recognition, **34**(9): 1751–1764.

Laptev, I., Lindeberg, T., Eckstein, W., Steger, C., and Baumgartner, A., 2000, Automatic extraction of roads from aerial images based on scale space and snakes, Machine Vision and Applications, **12**: 23–31.

Lei, H., Genxun, W., Changping, L., 2003, An improved parallel thinning algorithm, Proceedings of the Seventh International Conference on Document Analysis and Recognition (ICDAR 2003).

Mena, J.B., 2003, State of the art on automatic road extraction for GIS update: a novel classification, Pattern Recognition Letters, **24**: 3037–3058.

Pridmore, T., Bucha, V., Ablameyko, S., 2005, What should the user do? inference structures and line drawing interpretation, Document analysis and recognition, Proceedings of International Conference ICDAR, Seoul, pp. 760–764.

Smith, A., 1979, Tint fill, Computer Graphics, **13**: 276–283.

CLASSIFICATION OF REMOTELY SENSED DATA

SVITLANA KOKHAN
GIS Department, National Agricultural University of Ukraine, 15 Heroiv Oborony St., 03041, Kyiv, Ukraine

Abstract. The use of classifications for land cover mapping from satellite imagery is shown. Environmental and land cover maps represent the probable environmental statement of various types of land use and development of landscape. Remotely sensed data could be particularly efficient for environmental and land use mapping in order to outline main classification types.

Keywords: remote sensing, supervised classification, land cover mapping

1. Introduction

Remote sensing is the science (and to some extent, art) of acquiring information about the Earth's surface without actually being in contact with it. This is done by sensing and recording reflected or emitted energy and processing, analyzing, and applying that information. Remote sensing also involves the sensing of emitted energy and the use of nonimaging sensors.

The final element of the remote sensing process is achieved when we apply the information we have been able to extract from the imagery about the target in order to better understand it, reveal some new information, or assist in solving a particular problem.

Remote sensing is one of the major tools for the holistic approach to landscape evaluation, including satellite imagery (Kovalevskaya and Pavlov, 2002)]. The remote sensing satellites provide a combination of two types of information that can be used to assess landscape characteristics – the radiance of the earth's surface on a pixel-by-pixel basis and the spatial variability of radiance due to spatial patterns that could be detected. Spatial data consists of useful information to increase the potential of remote sensing in land use assessment. Spatial variability allows to derive data on land suitability and land use type, vegetation, and soil type.

A. Morris and S. Kokhan (eds.), Geographic Uncertainty in Environmental Security, 239–247.

Advances in technology are making global positioning systems, on the go yield monitors, GIS, remote sensing, communication networks, and variable rate application techniques available to producers. Scientists and farmers are using these technological approaches in agricultural research and precision farming.

Remotely sensed data and satellite imagery is an important input to many analyses. It can provide timely as well as historical information that may be impossible to obtain in any other way. The availability of this data provides opportunities for environmental studies particularly in the areas of change detection, land use mapping, land evaluation, land survey that would have been unknown of only a few decades ago.

2. Image Enhancement

Because remotely sensed imagery is a common source of data for GIS analyses and it has a raster structure, many raster geographic information systems provide some image processing capabilities. For the multispectral, landscape-scale phase of the study, we acquired cloud-free Landsat 7 scene (ETM+), made in July. We used Erdas Imagine V 9.0 (Leica Geosystems) to georeference image with ground-control points and IDRISI Kilimanjaro for image classification. Validation of classification was performed with field observations for all land use categories.

We explored different ways to increase the contrast of remotely sensed data to aid the image enhancement. The most simple type of stretch is a linear stretch using the minimum and maximum data values as the stretch endpoints. The endpoints of the data distribution are pulled to the endpoints of the palette and all values in between are rescaled accordingly (Eastman, 2003)].

To increase the contrast in the image we stretched the display so that all the colors of the palette, ranging from black to white were used. There were used two outcomes of stretch operations – the underlying data values remain unchanged and changes only to the display, and the creation of new image files. Two types of contrast stretch were used with images – linear stretches, with or without saturation, and histogram equalization (Figures 1, 2).

Better contrast can be achieved by applying a linear stretch with saturation to the image when we set new minimum and maximum display values that are within the original data value range. Most satellite images have distributions with narrow tails on one or both ends. While we use linear stretch with saturation all the values that lie above the new display maximum are assigned to the white color and all those below the new display minimum

Figure 1. Image enhancement

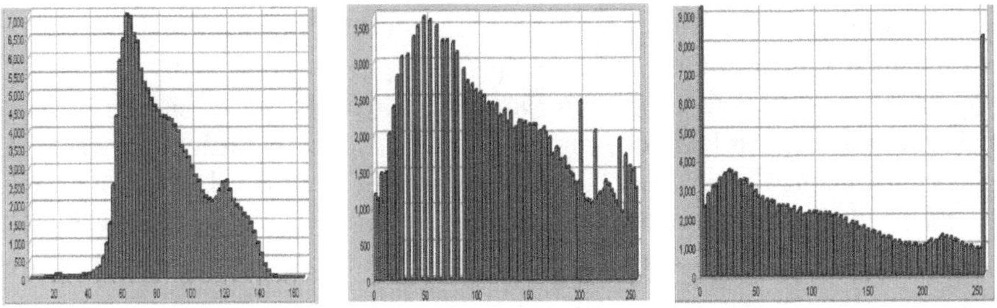

(A) histogram before the stretch

(B) histogram after the equalization

(C) histogram after the linear stretches with saturation (5%)

Figure 2. Histograms after the image enhancement

are assigned to the black color. The saturation of a user-specified percentage (e.g., 5%) of the pixels at each end (tail) of the distribution was applied to the image in order to enhance the contrast.

The histogram equalization stretch provides an image with very high contrast. Such an image can give more information than any other image because of the greatest variation for any given number of classes. The histogram equalization stretch assigns the same number of pixels to each data level in the output image. Pixels originally in the same category may not be divided into more than one category in the output image (Eastman, 2003)].

The creation of color composite images is one of the types of image enhancement. Color compositions give possibility to view the reflectance information from three separate bands in a single image.

3. Image Classification

Land cover and other kinds of maps may be developed from the classification of remotely sensed imagery. The majority of image classification is based on the detection of the spectral response patterns of land cover classes. Classification depends on distinctive signatures for the land cover classes in the band set being used, and the ability to reliably distinguish these signatures from other spectral response patterns that may be present (Eastman, 2003)].

The process of classification can be represented as one of the determining sets to which each pixel belongs. The sets in supervised classification assumed to be known before the process is begun. In the case of supervised classification we delineate specific land cover types based on statistical characterization data drawn from known as training sites.

Multiple schemes of image classification used in environmental mapping and monitoring as well as in agricultural management are either statistical (nonparametric rule of parallelepipeds; parametric rules of Maximum Likelihood, Minimum Distance, Mahalanobis Distance, etc.) or heuristic (labeling relaxation, region growing, fuzzy classification, etc.). A procedure is used to evaluate the likelihood that each pixel belongs to one of these classes is known as a classifier. These very specific examples within the image to be classified are called training sites because they are used to train the classifier on what to look for.

Traditional classifiers can be called hard classifiers because they yield a hard decision about the identity of each pixel. Soft classifiers express the degree to which a pixel belongs to each of the classes being considered.

A fuzzy interpretation of mapping of land cover replaces the traditional Boolean area-class map (Mark and Csillag, 1989)], which shows every pixel as belonging to one and only one class (Fisher et al., 2006)], by a set of n where n is the number of classes identified. For each pixel a membership value is given in the interval [0, 1]. The value 1 expresses a complete match between the characteristics of the location and those of the classes, and 0 indicates a complete mismatch (Fisher et al., 2006)]. The degree of matching has to be between 0 and 1.

Supervised classification foresees the identification of land cover types in the image. Examples of the information classes are called training sites. The next step in supervised classification includes the development a statistical characterization of the reflectance for each information class or signature development. Supervised classification requires that the user select training areas for use as the basis for classification. Various comparison methods are used to determine if a specific pixel qualifies as a class member.

In case of well-developed training sites the method of Maximum Likelihood is used. The Minimum Distance procedure with standardized distances is applied when the training sites are not uniform.

When the training sites are strongly representative of informational classes the Fisher Classifier can be used. In many cases the classified image has high comparability with a map.

Such classifiers as Parallelepiped, Minimum Distance to Means, and Maximum Likelihood are known to be hard classifiers because they make a definitive decision about the land cover class to which any pixel belongs. Application of these methods with Landsat 7 scene (ETM^+) images is shown in Figure 3.

(A) Minimum Distance to Means (B) Maximum Likelihood (C) Fisher Classifier

Figure 3. Hard classifiers

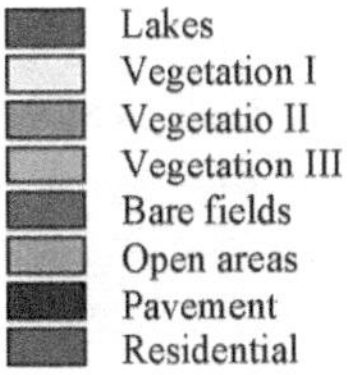

Because of a simple decision rule to classify multispectral data parallelepiped classification forms an n-dimensional parallelepiped in the image data space. Many pixel values in studied images fall in multiple classes and pixel is assigned to the last class matched. Therefore most of land cover types were represented by vegetation. Areas that do not fall within any of the parallelepipeds were designated as unclassified.

The Minimum Distance Classifier uses the mean vectors of each training site and calculates the Euclidean distance from each unknown pixel to the mean vector for each class. All pixels are classified to the nearest class unless a standard deviation or distance threshold is specified, in which case some pixels may be unclassified if they do not meet the selected criteria. The method gives some mistakes in classification results because of standard deviation of pixel spectral characteristics within the polygons (Figure 3A).

In case of Normalized Distance procedure application the classifier calculates standard deviation for reflectance values around the mean value and made contours of standard deviations. The pixel is assigned to the closest category in the form of standard deviations (Kokhan and Polishchuk, 2004).

The Maximum Likelihood procedure is one of the most sophisticated, and the most widely used classifier. The classification assumes that the statistics for each class in each band are normally distributed and calculates the probability that a given pixel belongs to a specific class. In Figure 3B there are eight land cover classes obtained using the Maximum Likelihood procedure. Each pixel is assigned to the class that has the highest probability.

So-called soft classifiers do not make a definitive decision about the land cover class to which each pixel belongs. They develop statements of the degree to which each pixel belongs to each of the land cover classes. For example in the image a soft classifier might indicate that a pixel has a 0.52 probability of being Residential, and 0.86 probability of being Open Areas (Figure 4). So, in case of a hard classifier application it would be resolved this value of uncertainty by concluding that the pixel was Residential. When we need to make a conclusion that the uncertainty increases because the pixel contains more than one cover type we could use the probabilities as indications of the relative proportion of each (subpixel level). The uncertainty growth could be connected with unrepresentative training site data. There is possibility to combine these probabilities with other evidence before hardening the decision to a final conclusion (Eastman, 2003).

The main soft classifiers and corresponding hardeners use the logic by which uncertainty is specified – Bayesian, Dempster-Shafer, and Fuzzy Sets (Porkhun, 2002) respectively.

Residential, Lakes, Vegetation (I through III) are the most common classes to which pixels show some significant degree of membership. There were small values for Bare Fields. This probably shows that the pixel either contains mixed cover classes or the reflectances of these pixels fall into overlapping sections of signature distributions of these classes.

The uncertainty for Residential at many locations is 0.50 and each cell has nearly equal membership in two classes. There is more uncertainty in this case because the two choices have similar support. In case of less uncertainty (Open Areas) one class has much more support than the other.

(A) Classification uncertainty (Bayesian probability theory)

(B) Classification uncertainty for Open Areas

Figure 4. Classification uncertainty

Vegetation land cover type represents some territories of very low uncertainty. Different types of vegetation can appear rather certain and these classes of vegetation do exist in large contiguous territories in the study area. The highest average uncertainty is for Roads class because of the possibility to be mixed with other cover classes for this image resolution (30 m) and Roads may often be mixed.

Dempster-Shafer theory is known to be a variant of Bayesian probability theory (Popov, 2002). The output of Dempster-Shafer classification is in the form of a series of belief images and a classification uncertainty image (Figure 5). The values in images represent the evaluated belief (a form of probability) that each pixel belongs to that class.

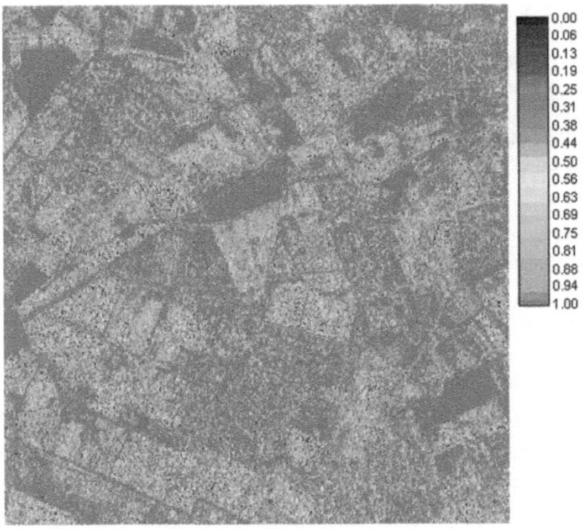

Figure 5. Classification uncertainty (Dempster-Shafer theory)

Results shows that the classification uncertainty is much higher when Dempster-Shafer theory is used compared to Bayesian probability theory. The Dempster-Shafer theory accepts the idea that unidentified cover types may exist. The beliefs of the Dempster-Shafer results have a mean value of 0.88 (compared to 0.23 for Bayesian probability theory application). Because Dempster-Shafer theory considers an additional unknown class – ignorance is introduced into the calculation and lowers the degree of certainty.

4. Accuracy Assessment

The final stage of the classification process usually involves an accuracy assessment. There are several types of accuracy assessment. Usually it is done by generating a random set of locations in the field conditions to verify of the true land cover type. Simple values file is then made to record the true land cover class for each of locations. This values file is then used with the vector file of point locations to create a raster image of the true classes found at the locations examined. This raster image is then compared to the classified map (Eastman, 2003).

The kappa coefficient is another measure of the accuracy of the classification. The coefficient is calculated by multiplying the total number of pixels in the ground truth classes by the sum of the confusion matrix diagonals, subtracting the sum of the ground truth pixels in a class times the sum of the classified pixels in that class summed over all classes, and dividing by the total number of pixels squared minus the sum of the ground truth pixels in that class times the sum of the classified pixels in that class summed over all classes. The overall Kappa for Landsat image classification using the cross-tabulation is 0.69.

The error matrix produced may be used to identify particular cover types for which errors are in excess of that desired. The information in the matrix about which covers are being mistakenly included in a particular class (errors of commission) and those that are being mistakenly excluded (errors of omission) from that class can be used to refine the classification approach.

5. Conclusions

The principles of supervised classification can be applied on a different scales. Many users have explored fuzzy classification of land cover, but the purpose of the paper is to show the application of two others soft classifiers – the Bayesian probability theory and Dempster-Shafer theory.

Bayesian analysis, by assuming complete knowledge is present, produces almost equally high probabilities for spatial categories that may have similar or overlapping reflectance patterns while Dempster-Shafer theory recognizes these areas are different.

Because Dempster-Shafer theory considers an additional unknown class – ignorance is introduced into the calculation and lowers the degree of certainty.

References

Eastman, J. R., 2003. Guide to GIS and Image Processing. Clark University Manual Version 14.00.

Fisher, P., Arnot, C., Wadsworth, R., Wellens, J., 2006. Detecting change in vague interpretations of landscapes. Ecological Informatics 1, 163–178.

Kokhan, S., Polishchuk, I., 2004. Remote sensing monitoring for land resources. Kiyv, NAUU, 68p.

Kovalevskaya, N., Pavlov, V., 2002. Environmental mapping based on spatial variability. Journal of Environmental Quality 31, 1462–1470.

Mark, D.M., Csillag, F., 1989. The nature of boundaries on 'Area-Class' maps. Cartographica 26, 65–78.

Popov, M., 2002. Present-day views on the interpretation of RSE data. Space Science and Technology, V. 8, 2/3, 110–115.

Porkhun, O., 2002. Application of geoinformation systems to the interpretation of aerospace images. Space Science And Technology, V.8, 2/3, 106–109.

SUSTAINABILITY AND ENVIRONMENTAL SECURITY MANAGEMENT TOOLS

ALEXANDER GOROBETS*
Sevastopol National Technical University, Management Department, Streletskaya Bay, Sevastopol 99053, Ukraine

Abstract. In this paper, the problem of sustainable development is highlighted on the global level and for particular country (Ukraine). Strong and weak approaches to sustainable development and specific management tools, i.e., economic, institutional, social (education), and technological (GIS) are given to address environmental security issues. The integrated characteristic of environmental and human well-being (health) is proposed as the principal indicator of sustainable development.

Keywords: sustainable development, health, ecological economics, management tools

1. Introduction

"Sustainable Development seeks to meet the needs of the present generation without compromising the ability of the future generation to meet their own needs." With these words, in 1987 the World Commission on Environment and Development (the Brundtland Commission) put the concept of sustainable development on the international agenda. Sustainable development combines concern for economic progress and the elimination of poverty with awareness of environmental limits. It raises questions of population, affluence, consumption equality and the resource, energy and pollution intensity of production and technology. However, in spite of many organizations working in this field, international intergovernmental meetings and official declarations, vital for humanity problems of climate change (UNEP, 2006), biodiversity loss (WWF, 2004), environmental pollution and wastes

* Alexander Gorobets, Sevastopol National Technical University, Management Department, Streletskaya Bay, Sevastopol 99053,Ukraine; alex-gorobets@mail.ru

A. Morris and S. Kokhan (eds.), Geographic Uncertainty in Environmental Security, 249–257.

accumulation (e.g., nuclear) have very dangerous systematically stable trends. The main reasons for that are as follows:

1. Extensive type of economic development and poor understanding of sustainable development principles both by the government authorities and by public due to a lack of qualified staffing potential
2. Weak decision-making and management on all levels and psychological (socio-cultural) and institutional barriers (inertia) of taking measures in a new, rapidly changing and complex environment
3. Absence of the clear consistent goals and specific well-developed national programs of sustainable development and contradictions between existing indicators

Therefore the purpose of this paper is to develop the integrated consistent indicator of sustainable development and appropriate management tools that will be consonant with the national strategies of sustainable development and millennium development goals: poverty reduction, quality life-long education, environmental sustainability, improved health and reduced HIV/AIDS and other diseases, gender equality, global partnership.

This paper proceeds as follows. In the second section the problem of sustainable development is considered on the global scale. Third section puts emphasis on sustainable development in Ukraine and proposes a new integrated characteristic of sustainable development. Fourth section presents different approaches to sustainable development and specific management tools and finally the paper closes with conclusions.

2. The Problem of Sustainable Development on the Global Scale

The impact of human activity on environment can be defined as follows:

$$I = PAT, \tag{1}$$

where P is population, A is affluence (wealth) and T is technological factor.

The dynamics of first two factors is shown on the Figures 1 and 2 respectively (UNEP, 2006). Both of them have stable upward trends in the last decades and there is no evidence that they can stabilize in the nearest future. In the same time, development of technologies allows to reduce energy and material intensity of economic activity and therefore compensate (at least to some theoretical limits) the influence of growing population and its material wealth.

Indeed, such indicator as energy supply per US \$1,000 of gross domestic product in purchasing power parity is improving due to technological progress in the most of the regions (UNEP, 2006).

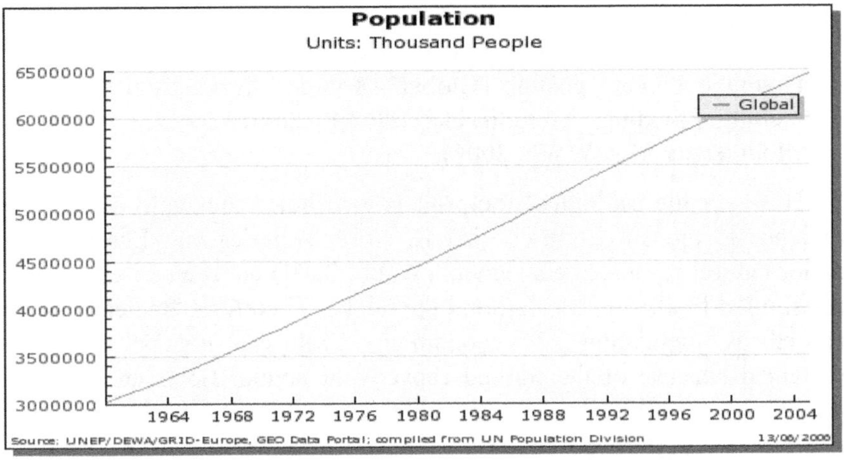

Figure 1. Dynamics of the global population in the last decades

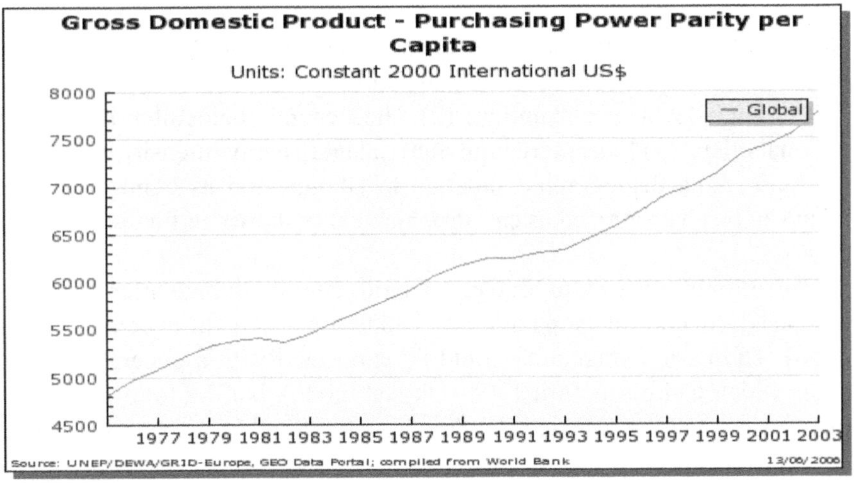

Figure 2. Dynamics of the global gross domestic product in purchasing power parity per capita in USD

However, up-to-date technological progress is going slower than a rise of population and its affluence that leads to increasing pressure on environment. It is confirmed by such indicator as ecological footprint – the amount of land water area a human population would need to provide the resources required to sustainably support itself and to absorb its wastes, given prevailing technology, which is growing (WWF, 2004). According to this indicator, "material growth"-oriented humans now consume (deplete) the natural capital at the rate above the carrying capacity of Earth, causing severe environmental threats to present and future generations simultaneously with:

- High growth of municipal wastes, emissions of air, water pollutants (Affecting human health) and carbon dioxide (UNEP, 2006)
- Dramatic climate change (Global warming, Agricultural change, Ice melting, Flooding, Droughts, etc.) (UNEP, 2006)
- Biodiversity loss (WWF, 2004)

However the ecological footprint is distributed unequally in the world regions, developed countries, that is, North America and Europe are the major natural resources consumers (WWF, 2004) but the consequences are transmitted to all countries around the Globe. Therefore, "material growth" caused by population and economic growth can not be ecologically sustained because of the limited capacity of natural flows and cycles for providing the resources and absorbing or assimilating the waste (such as heavy metals or carbon dioxide).

3. The Problem of Ukraine' Sustainable Development

Additionally to the obstacles of sustainable development that are common for all countries (see section 1), Ukrainian nation has the following problems:

(1) High level of corruption; (2) The general sociocultural and moral national crisis; (3) Extremely high material and energy intensity economy.

As a result, demographic situation in Ukraine and its trend of development can be characterized as catastrophic, that is shown on Figure 3 (Ukrstat, 2006).

According to the State Statistics Committee of Ukraine after 1992 there is a steady decline in population at 388,000 per year in average (Ukrstat, 2006). The basic explanation could be connected with a general economic crisis which had place until 1998 (Ukrstat, 2006), but the following economic growth has not affected the demographic trend because this economic

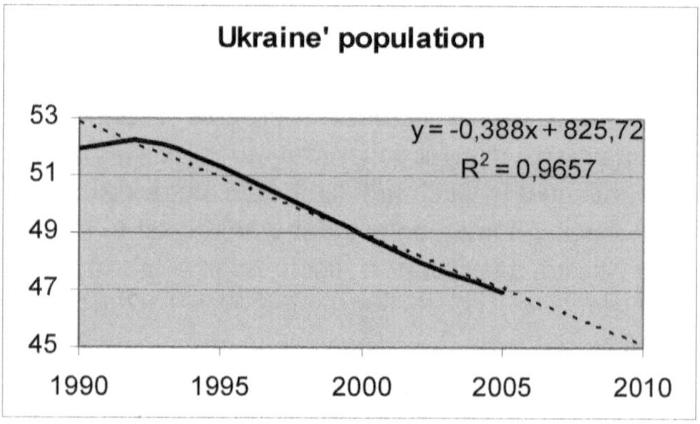

Figure 3. Demographic situation in Ukraine and its trend of development

benefit was not used efficiently (e.g., by investing in health care and human development system) and distributed equally among different groups of population, that is, the gap between rich and poor has been increasing. According to the estimated trend equation (fitted well to a given statistics – Figure 3) Ukraine' population will decrease to about 45 million by 2010 if the present inhuman socioeconomic policy is not changed. This trend is based mostly on a natural population decrease (number of deaths is more than number of births), where the most of deaths is caused by blood circulation disease (Ukrstat, 2006).

Therefore, sustainable development in Ukraine and many other transitional, developing, and developed countries should be based on both the intergenerational and intragenerational socioeconomic and ecological justice of society (Pearce, 1987) and focused on a permanent livability (adoption of eco-centric rationale instead of dominating anthropocentric one) and ecosystems and human well-being (Dodds, 1997):

- The necessary material minimum
- Freedom and choice
- Health and bodily well-being
- Environmental well-being
- Good social relations
- Personal security and
- The conditions for physical, social, psychological, and spiritual fulfillment.

The Human Development Index (HDI) is often considered among the integral indexes of sustainable development to measure and compare the progress in different countries. The index is elaborated within the framework of the United Nations Development Program (UNDP, 2005). The HDI consists of three interrelated components:

(1) Health, measured by life expectancy at birth
(2) Education – adult literacy rate (2/3 weight) and the combined primary, secondary and tertiary gross enrolment ratio (1/3 weight)
(3) Standard of living – Gross Domestic Product per capita at Purchasing Power Parity in USD

However, the HDI has some weak points. It concentrates on the quantitative parameters of human life according to the principle "more is better" (especially for GDP per capita) rather than on its qualitative growth. Clearly human health might be considered as the biggest value but it should not be measured only by life expectancy (especially in developed countries with strong medical system) but also by sickness rate and physical and intellectual capacity. Furthermore, education and standard of living have direct impact on health, especially in developing countries, so all three components are strongly correlated and can be replaced by one integrated characteristic of human health which is a state of complete physical, mental, and social well-being.

Human health depends not only on human biology (genetic peculiarities), the public health system, but also is concerned with the state of the economy, the social, physical, spiritual, and moral life (lifestyle) and the state of the environment. According to the World Health Organization, environmental risk factors play a role in more than 80% of the diseases (Prüss-Üstün and Corvalán, 2006).

Therefore, in this paper, the integrated characteristic of human health (physical and intellectual capacity, life interval, sickness rate, psychological health) is proposed as the principal indicator of sustainable development.

4. Approaches to Sustainable Development and Management Tools

Although there are many complementary approaches to sustainable development, in this paper the emphasis is given to the most different approaches, i.e., strong sustainability and weak sustainability (Costanza, 1991). Weak sustainability (anthropocentric approach) assumes that natural and manmade capital are substitutes and requires that their sum is nondeclining. Strong sustainability (eco-centric approach) assumes that, at least in some important aspects (e.g., ecosystem services), substitution is not possible and requires nondeclining stocks of natural capital. Environmental economics (part of neoclassical economics concentrating on material growth) is a theoretical base for weak sustainability, while ecological economics (Costanza, 1996; Costanza et al., 1997), based on the fundamental thermo dynamical laws (governing the natural ecosystems) contributes to strong sustainability. The major differences between ecological economics and environmental economics are given in Table 1 (Bergh and van den, 2000).

TABLE 1. Differences between ecological economics and environmental economics

Ecological economics	Environmental economics
1. Optimal scale, growth pessimism	1. Optimal allocation, growth optimism
2. Priority to sustainability	2. Priority to efficiency
3. Human needs and fair distribution	3. Optimal welfare, Pareto efficiency
4. Long-term focus	4. Short to medium term focus
5. Systems analysis	5. Mono disciplinary and analytical
6. Concrete and specific	6. Abstract and general
7. Physical and biological indicators	7. Monetary indicators
8. Bounded rationality	8. Maximization of utility
9. Local communities	9. Global market
10. Environmental ethics	10. Utilitarianism

To achieve sustainable development goal (health), specific management tools, i.e., economic, institutional, social, and technological are presented in Table 2.

One of the most important among these tools is Geographic Information Systems (GIS), which are powerful instrument in collecting and analyzing information about geographic objects with uncertain boundaries for environmental security and protection. The ultimate use of GIS lies in its capability for modeling cause and effect scenarios, identifying trends and factors that affect them and prediction the consequences of natural resources use and management (FAO, 2006).

Application of GIS technologies in Ukraine is very relevant and important, especially in agricultural sector (selection of the best potential sites for specific crops, fishing management), forest industry, nature protection (illegal chopping, hunting, fishing, e.g., in the Black and Azov Seas), environmental security (air and water industrial and municipal pollution). The present state of Ukraine' environment is fragile indeed because of the many borders (with seven countries), long coastal zone, climate change and extensive type of economic development, sociocultural crisis and irresponsible and incompetent environmental management on all levels.

However, application of GIS in Ukraine has to be accompanied by the improvement of existing information collection and processing systems and the introduction of new ones.

TABLE 2. Approaches to sustainable development and management tools

Approach	Scale	Management tools
Strong sustainability ecological economics	Microlevel communities	Precautionary and responsibility principles, Education for Sustainable Development (UNESCO, 2003) – awareness rising, changing attitudes and values (Maiteny, 2000), capacity building, Communities empowerment, institutions, support of local knowledge and community identity, better indicators of human well-being (Daly and Cobb, 1994)
Weak sustainability environmental economics	Macrolevel	Cost–benefit analysis, industrial ecology, life cycle assessment, eco-efficiency, eco-taxation, economic valuation of environmental services (Brown, 2001), environmental risk management, green accounting (UNSTAT, 2006), ISO 14000, GIS – mapping of resources, uncertainty modeling (Langaas, 1997, FAO, 2006)

5. Conclusions

In this paper, the integrated characteristic of human health, i.e., physical and intellectual capacity, life interval, sickness rate, psychological health is proposed as the principal indicator of sustainable development. To achieve sustainability goal, specific management tools are proposed. The capacity of geographic information systems and its prospective application are shown for Ukrainian case.

The advantage of developed management tools is that they can significantly contribute to environmental security in all countries, especially in transition and developing countries, although every country and local community around the Globe must find their own way of sustainable living depending on its local environment, cultural traditions and socioeconomic development.

Transition to sustainability in Ukraine is a vital issue due to the critical state of its population, environment, and uncertainty of economic development. Only deep and quick consolidation of all institutions (governmental, educational, etc.) is able to make difference in sociocultural, economic, and environmental development through the strict (and correct) regulation and systemic education and upbringing of whole population to achieve internal human sustainability grounds (Gorobets, 2006).

References

Bergh, J.C., van den, J.M., 2000, Ecological economics: themes, approaches, and differences with environmental Economics, Tinbergen Institute Discussion Paper TI 2000-080/3; http://www.tinbergen.nl.

Brown, L.R., 2001, Eco-Economy: Building an Economy for the Earth. W.W. Norton & Company, NYk: 352p; http://www.earth-policy.org/Books/Eco_contents.htm.

Costanza, R., (ed.), 1991, Ecological Economics: The Science and Management of Sustainability. Columbia University Press, NY.

Costanza, R., 1996, Ecological economics: reintegrating the study of humans and nature, Ecological Applications **6**(4): 978–990.

Costanza, R., Cleveland, C. and Perrings, C., (eds.), 1997, The Development of Ecological Economics, E. Elgar, Cheltenham, UK.

Daly, H., and Cobb, J., 1994, For the Common Good: Redirecting the Economy toward Community, the Environment and a Sustainable Future, 2nd edn. Beacon, Boston, MA.

Dodds, S., 1997, Towards a 'science of sustainability': improving the way ecological economics understands human well-being, Ecological Economics **23**(2): 95–111.

FAO, 2006, Geographic information systems in sustainable development. Food and Agriculture Organization of the United Nations; http://www.fao.org/.

Gorobets, A., 2006, An eco-centric approach to sustainable community development, Community Development Journal **41**(1): 104–108.

Langaas, S., 1997, The spatial dimension of indicators of sustainable development: the role of Geographic Information Systems (GIS) and cartography, in: B. Moldan and S.

Billharz (eds.), Sustainability Indicators: A Report on the Project on Indicators of Sustainable Development. SCOPE 58, John Wiley & Sons, Chichester, 33–39.

Maiteny, P., 2000, The psychodynamics of meaning and action for a sustainable future, Futures **32**: 339–360.

Pearce, D., 1987, Foundations of an ecological economics, Ecological Modelling, **38**: 9–18.

Prüss-Üstün, A., and Corvalán, C., 2006, Preventing disease through healthy environments: towards an estimate of the environmental burden of disease: executive summary. World Health Organization, **WA 30.5**. Geneva; http://www.who.int/en/.

Ukrstat, 2006, State Statistics Committee of Ukraine, Kiev; http://www.ukrstat.gov.ua/.

UNDP, 2005, Human Development Report 2005, Human Development Index. UNDP, New York; http://hdr.undp.org.

UNEP, 2006, The GEO Data Portal. United Nations Environment Programme; http://geodata.grid.unep.ch.

UNESCO, 2003, United Nations Decade of Education for Sustainable Development (2005–2014). Framework for the international implementation scheme, UNESCO **32 C/INF.9**. Paris; http://unesdoc.unesco.org/images/0013/001311/131163e.pdf.

UNSTAT, 2006, Handbook of National Accounting: Integrated Environmental and Economic Accounting – An Operational Manual, UNSTAT **F, No.78**. New York; http://unstats.un.org.

WWF, 2004, Living Planet Report 2004. WWF International, Gland; http://www.panda.org.

REMOTE SENSING AND GIS APPLICATION FOR ENVIRONMENTAL MONITORING AND ACCIDENTS CONTROL IN UKRAINE

D.K. MOZGOVIY, O.I. PARSHYNA, V.I. VOLOSHYN, Y.I. BUSHUYEV
State Company "Dniprocosmos" of National Space Agency of Ukraine

Abstract. Current status of implementation of Earth remote sensing technologies in Ukraine, structure of the Ukrainian Segment of GEOSS-GMES, some examples of works for solving of nature management problems are offered in article.

Keywords: remote sensing data, geospatial information, GEOSS-GMES-Ukraine

1. Current Status of Implementation of Earth Remote Sensing Technologies

Earth remote sensing data is the most important source of geospatial information due to high visibility, responsiveness, multidisciplinarity and objectivity.

By preliminary estimates, almost 80% of environment indicators (indices) can be determined using Earth remote sensing data (Voloshyn et al., 2005a).

Aerospace information due to its objective nature is used as an element that supplements, generalizes, and details information obtained from conventional terrestrial sources. Thus Earth remote sensing data serve as the *basic procedure for verification of data obtained from various departmental sources*. Therefore wide implementation of Earth remote sensing technologies is essentially *a structural reorganization of the state geoinformational provision and belongs to nation-wide problems by its scale and expected effect*.

Urgency and technical–economical expediency of wide implementation of Earth remote sensing technologies are well realized at world and European levels, which is characterized by realization of appropriate ambitious projects – GEOSS, GMES, INSPIRE – in the first years of 21st century.

A. Morris and S. Kokhan (eds.), Geographic Uncertainty in Environmental Security, 259–270.

The main obstacles on the way of implementation of Earth remote sensing technologies in everyday activity of subjects of economical, scientific, and management efforts are the series of factors of organizational, normative-legal, methodical, informational, and technical nature. The most important of these factors are the following ones:

Earth remote sensing data is only electromagnetic images of discovered surface with indirect information about physical phenomena and processes, which in fact, interest an end user (agronomist, forester, cartographer, builder, etc.). To obtain information useful for an user it is necessary carrying out sufficiently time-consuming operations of thematic decoding of an image with utilization of auxiliary terrestrial data.

A user needs information "at the specific time and about the necessary region". However currently existing technology of space imaging scheduling and performing not always enables fulfilling this requirement. For optical imaging this situation is additionally complicated by cloud canopy condition that is forecasted with difficulty. Thus for successful solving of wide class of operative applied problems, the fundamental reorganization of technology of space imaging scheduling and space data obtaining is necessary.

There are no certified procedures of Earth remote sensing data processing, which would provide obtaining of final information products with guaranteed quality.

There is no normative-legal base, which would regulate rules of Earth remote sensing data exchange, processing, and use.

Most of users have not economic interest in implementation of technologies of thematic problems solving with utilization of space information because of considerable initial investments and long-term (2–5 years) realization. At the same time there are no organizational structures, which would integrate interests of different users and might be an intermediate party at solving this problem.

Most of space remote sensing information, which is used by Earth sciences, for weather and climate forecasting, for ecological monitoring, etc., generally has not commercial value, and its economical effectiveness is estimated indirectly, for instance, by level of national security, by results of fundamental scientific researches, by extent of excluded damage, and so on.

Complicated access to current and archive data of Earth remote sensing, to necessary auxiliary terrestrial data, to map materials because of institutional, customs, security, financial barriers. Furthermore national satellites of Earth remote sensing do not provide continuous series of observations that is necessary for practice.

Insufficient number of experts on remote sensing data processing and decoding with use of modern computer facilities is also the critical factor in widespread implementation of Earth remote sensing technologies.

In the project of Space program of Ukraine designed by National Space Agency of Ukraine (NSAU) for 2007–2011, creation of Ukrainian national segment of environmental monitoring and accident prevention – GEOSS–GMES–Ukraine – is stipulated [Lyalko and Popov, 2006].

2. Structure of the Ukrainian Segment of GEOSS-GMES

Ukraine has the following necessary components of space infrastructure Space segment – "Sich" Earth observation system:

Ground information complex for reception, preliminary processing, and archiving and distribution data of Earth remote sensing from national and foreign satellites.

These components are the primary source of Earth remote sensing data but they do not eliminate the above listed obstacles for the data utilization.

To provide effective utilization of aerospace Earth remote sensing data and to support users' activities NSAU has initiated development of the *NSAU CosmoGIS* departmental information system in 2004.

Purpose of NSAU CosmoGIS is provision of users' geographic information systems (GIS) by Earth remote sensing data and services; it is a part of the state monitoring system. NSAU CosmoGIS supplements existing and

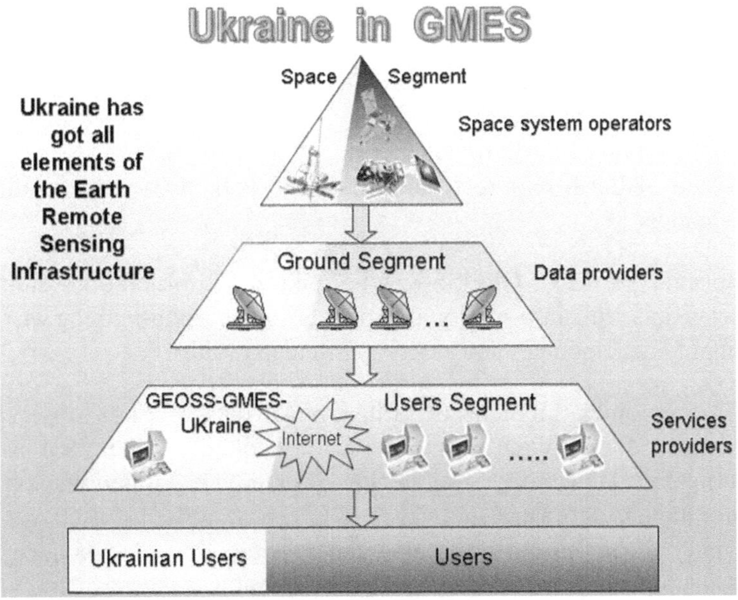

Figure 1. Ukraine in GMES

developing departmental information analytical systems (IAS) and monitoring services that use Earth remote sensing data and services.

Interaction with these systems is implemented via organization of corresponding innovation structures (consortiums) at the specific thematic directions together with the users profile institutions.

Pursuant to the project of the new Space Program of Ukraine for 2007–2011. [http://www.nkau.gov.ua/nsau/newsnsau.nsf/NewsLastE/FFA3B5BAC04153FCC225718000515BD4?OpenDocument&Lang=E] CosmoGIS will develop as an *interdepartmental system* for introduction of remote sensing technologies in various spheres of public activities and as the national contribution to construction of systems *GEOSS–GMES*.

In this connection the system since 2007 will receive new name: *GEOSS–GMES–Ukraine*.

GEOSS–GMES–Ukraine information system is not an alternative or modification of existing departmental information analytical systems. Purpose of its creation provides for expansion of information capabilities of each the existing systems and decision making centers on the basis of utilization of Earth space observation data, methods of their processing, and archiving as well as expansion of interdepartmental and international collaboration in this area.

Current activity and responsibility for status of geoinformation provision is owed as prerogative of subjects of state environmental monitoring and accident prevention in accordance with their scope of activity under the existing legislation.

In this sense the GEOSS–GMES–Ukraine system plays in national infrastructure of spatial data role of an innovative structure, which provides:

Implementation of Earth remote sensing data utilization

Provision of Earth remote sensing data and their processing results on users' demands

The primary tasks of GEOSS–GMES–Ukraine information system are users' provision with informational products for decision-making in fields of sustainable development and safety, particularly with:

GIS technologies, methods of utilization of data of Earth observation from satellites for solving of problems of ecological and food safety, effective nature management, monitoring of emergency situations, agriculture, water industry, forestry, etc.

Estimations of hazards of emergency situations' appearance and recommendations relative to their risk management

Prediction calculations and cartographical materials, which reflect global and regional scenarios of climatic, ecological, and socioeconomic changes and progress trends of the country territory

Analysis and prediction of behavior of ecosystems' parameters in conditions of anthropogenic loads buildup

Recommendations relative to enhancement of system of rational nature management

The main method of performing of the formulated tasks is integration of information capabilities of Earth remote sensing and capabilities of existing departmental structures of environmental monitoring and investigation within the common approach. Model of the integration is realized by the way of creation and provision of operation of an interdepartmental structure including:

GEOSS–GMES–Ukraine system central part or core as a network of Earth remote sensing service centers

Thematic segments: Emergency Situations, Health, Energy, Climat, Water, Ecosystems, Biodiversity, Safety, and so on

Technological segments, which provide the system with Earth remote sensing data and basic land data in general use

Earth remote sensing service centers within the GEOSS–GMES–Ukraine system are intended for users' information support, for users' provision with package of normative documents, which regulate activity for obtaining, processing, and utilization of Earth remote sensing data, for methodological support of Earth remote sensing data processing as well as for ensuring of international collaboration of Ukraine in field of space engineering. Their main destination is development and implementation of technology of thematic problems solving with utilization of Earth remote sensing data.

Institutions of National Space Agency of Ukraine (NSAU) and organizations of dual subordination to National Space Agency of Ukraine and National Academy of Sciences of Ukraine (NSAU&NASU), which have necessary experience and equipment for creation and implementation of technologies of utilization of Earth remote sensing data will perform function of the service centers (Figure 2). Among these are:

Dniprocosmos State Company of NSAU and Dnipropetrovsk Oblast State Administration – GEOSS–GMES–Ukraine project coordinator

National center of space facilities control and testing of NSAU (NCSFCT)

Space Research Institute of National Academy of Sciences of Ukraine and National Space Agency of Ukraine (SRI NASU–NSAU)

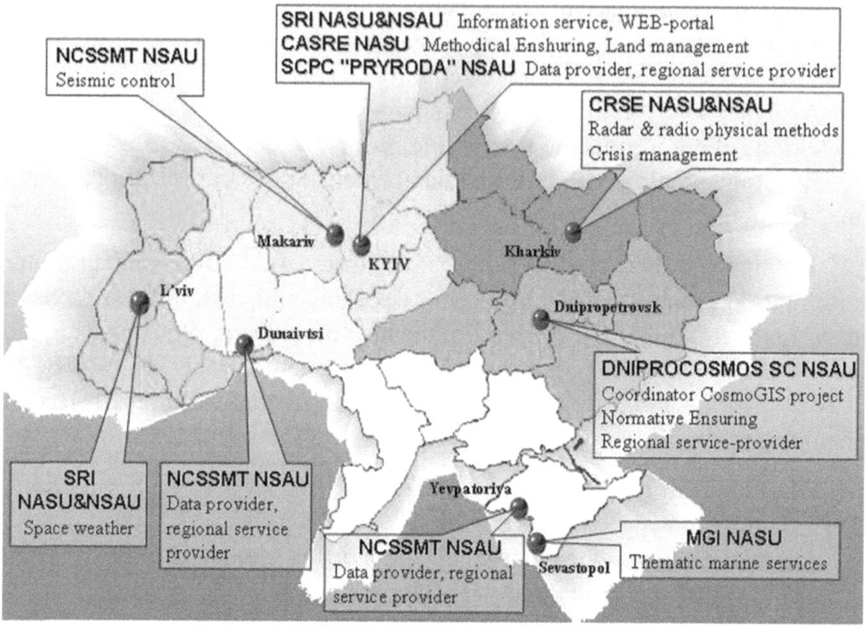

Figure 2. Service centers of GEOSS–GMES–Ukraine information system

Centre for Aerospace Research of the Earth of Institute of Geological Sciences of National Academy of Sciences of Ukraine (CASRE) [3]

Marine Hydrophysical Institute of National Academy of Sciences of Ukraine (MHI NASU)

Kalmykov Center for Radiophysical Sensing of the Earth of National Academy of Sciences and National Space Agency of Ukraine (CRSE NASU–NSAU)

State Scientific Production Center of Aerospace Information, Earth Remote Sensing and Environmental Monitoring “Pryroda” (SSPC “Pryroda”)

Thematic segments are the main facility for integration of satellite and land data within the common problem-oriented approach. They perform the primary interface role between Earth remote sensing data providers, land data providers from various departmental sources, satellite and land data processing centers and end users of the information product.

The main functions of the thematic segments are:

Development and implementation of technology of solving of the specific thematic problem or the problems logical collection with utilization of Earth remote sensing data

Provision of users with end information product in accordance with the technology

Technological segments provide data reception from Earth observation satellites, preprocessing, archiving, and distribution of Earth remote sensing data, ensuring of users' interactive access to catalog of archive of Earth remote sensing data, imaging performing with attraction of laboratory aircraft, and other technological operations connected with obtaining and providing data for service centers and thematic users.

Institutions that ensure the system with basic sets of spatial data in general use are also among the technological segments.

3. Brief Description of the Project Coordinator

Primary intent of Dniprocosmos State Company is implementation of aerospace technologies in practice of economic and management activity.

During the years of its activity, company has accomplished a number of projects:

Development and put into experimental operation regional complex of aerospace monitoring and program-technical complex for processing of aerospace images (1998–2002, NSAU project)

Development and realization in 1998–2002 regional program of data of Earth space observations for solving problems of subjects of management and economical activity in Pridneprovj'e region (1998–2002, project of Dnipropetrovsk Oblast State Administration and NSAU)

Experimental project on control of location and state of particularly dangerous mobile, hard-to-reach and remote objects using satellite means of navigation, communication, and transfer of data to control center (2001–2005, NSAU project)

Development of automatic system for ecological monitoring of city of Dnipropetrovsk (2002–2005, project of Dnipropetrovsk Oblast State Administration)

Documentation for users of "Sich-1" and "Okean-O" space systems (1994–2000, NSAU project)

The state standard of Ukraine 4220-2003 "Remote sensing of Earth from space. Terms and definitions" (2002–2004, NSAU project)

Development of software for Sich1-M Mission Control Center

Development of five programming-technical complexes for image processing that are transferred to other organizations to operation (2000–2005); on IDL base about 15 program modules are developed and integrated into ENVI environment

Dniprocosmos State Company has experience in Earth remote sensing data processing obtained from national spacecrafts «Sich-1M», «Ocean-O»,

«Sich-1», as well as from foreign ones – «Meteor-3M», Terra (Modis, Aster), IRS, Ikonos, Landsat, Spot.

Works for solving of some nature management problems were performed by the company, including: determination of man-caused load on vegetation of Dnipropetrovsk city, dynamic of planting of greenery in the city, pollution of Dnieper river within the precincts of the city by suspended substances, determination of pollution from industrial dumping in Dnieper, mapping of landslide emergency zones in territory of the city, determination of changes of urban development, mapping of soils in some farms of the region, determination of winter wheat sprouts area in administrative districts and the region as a whole, etc. Some examples are presented in Figures 3 and 4.

The company also performs development of proprietary methods and algorithms of data processing. For example, one of these engineering is a method for radiometric correction of multispectral images to ensure enhancement of information content and spatial resolution. The method [4] offers correction in the case of point displacement between different spectral channels due to diffraction effects and spatial coordinatesusceptibility of photoreceivers causing positioning instability during sensing. As applications of this method, some case studies on potentially dangerous landslide areas selection and classification of land cover elements classification have been developed as well (Figure 5).

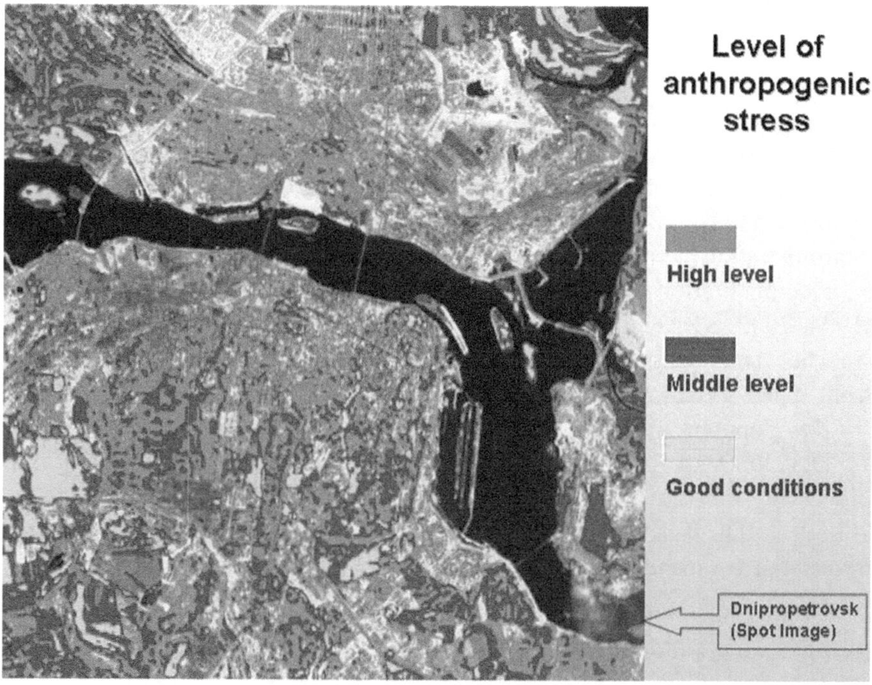

Figure 3. Image fragment and the circuit of density lineaments by results of decoding

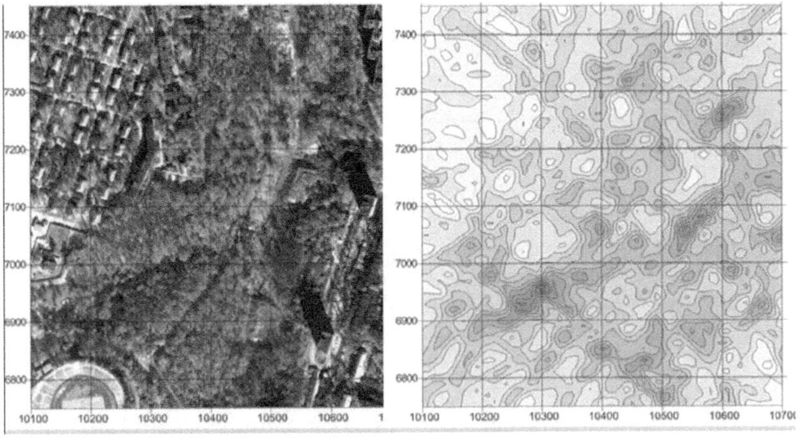

Figure 4. Mapping in the administrative district

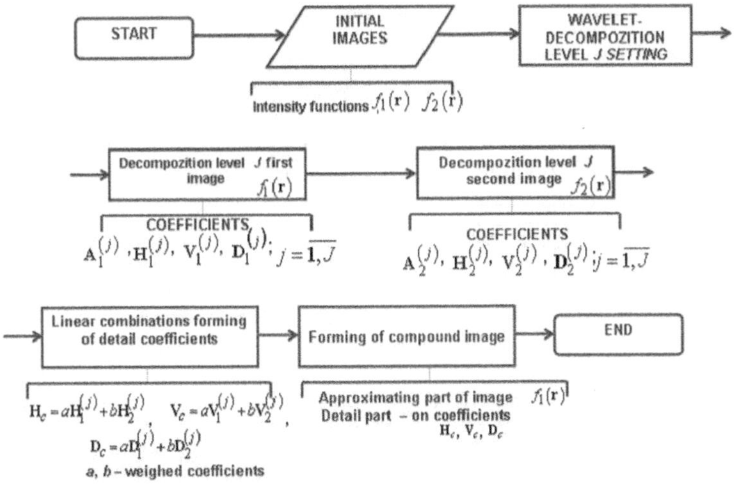

Figure 5. A block diagram of the overall method for radiometric correction of multispectral images to ensure enhancement of information content and spatial resolution

The main current projects performed by Dniprocosmos State Company are:

Creation of GEOSS–GMES–Ukraine system.

Development of State Standards in field of Earth remote sensing:

SSU 4220 Remote Sensing of the Earth From Space. Terms and Definitions

SSU Data processing. Terms and Definitions

SSU Data Processing Strategy. General Regulations

SOU NSAU Data Quality Indexes. General Requirements

SOU NSAU Data Processing Strategy Classifier

SOU NSAU Methods of Users Needs Assessment in Remote Sensing Data. General Regulations

Development and certification of methods of Earth remote sensing data processing:

Forests composition and quality assessment

Black and Azov Sea water temperature mapping on the base of AVHRR scanner

Rapid observing of the cyclones, frontal zones and whirlpools on sea surface

Ice-Water verge observing and control

Scanner data atmosphere correction

Land cover classification

Methods oriented on temporary and accepted information sources

Methods connected with NewEUser project and oriented on next thematic services:

- *Land management*
- *Crisis management*
- *Marine services*

Implementation of "Agrocosmos" pilot thematic segment for monitoring of agrarian resources of the country with utilization of Earth remote sensing data. This project is performed in collaboration with Agricultural biology institute and other institutions of Ukrainian academy of agrarian sciences.

Priority measures of the GEOSS–GMES–Ukraine project are the following:

Development of service centers' network ensures internet access to Earth remote sensing resources.

Creation of distributed databases of Earth remote sensing data of high (better than 30 m) and ultra-high (better than 5 m) spatial resolution and keeping them in actual status. For solving of this problem ensuring of data reception from foreign satellites onto national data reception station is keep an actual task.

Creation of pilot thematic segment in field that are actual for Ukraine – Agriculture (Agrocosmos), Marine ecosystems (Aquacosmos), Emergency situations.

Creation, finalizing, and certification of methods of Earth remote sensing data processing for obtaining of basic production for general multipurpose use (space maps, orthospace maps, thematic maps of land covers, temperature abnormalities, snow cover, marine surface temperature, etc.). This problem plays the key role in formation of National infrastructure of spatial data (NISD) and ensuring of constructive interaction with existing departmental systems of environmental monitoring as well as in NISD harmonization with European infrastructure of spatial data. INSPIRE (The INfrastructure for SPatial InfoRmation in the Europe) project [Inspire, 2002, Voloshyn et al., 2005b] that is currently performed by initiative of European Commission and European Environmental Agency (EEA) determines the common for all the European countries structure of spatial environmental information, rules of data obtaining, processing, exchange, and distribution. For Ukraine the common unified approach to the data, formation of system of environmental indices, which can be obtained with use of Earth remote sensing data is also important.

Draft method for land covers classification developed by Dniprocosmos State Company [Voloshyn et al., 2004] as a whole corresponds to All-European specification (CLC and CLC-2000 projects [Büttner, 1998-CORINE 2000]) in part of requirements to map material's scale, to characteristics of classification accuracy, to decomposition on thematic classes, but requires serious revisions with the purpose of automation of data processing and increasing of classification accuracy (Figure 6).

Our company plans to do the following works in this direction:

Harmonization with INSPIRE directives as well as the NIS governmental, national, and regional directives will be ensured.

Expert knowledge about feature signatures (spectral, textural, contextual) of separate classes will be formalized in form of definition, processing algorithms, structures, and rules.

Substantiation Informative Decoding Features (IDF) for each Land Cover class.

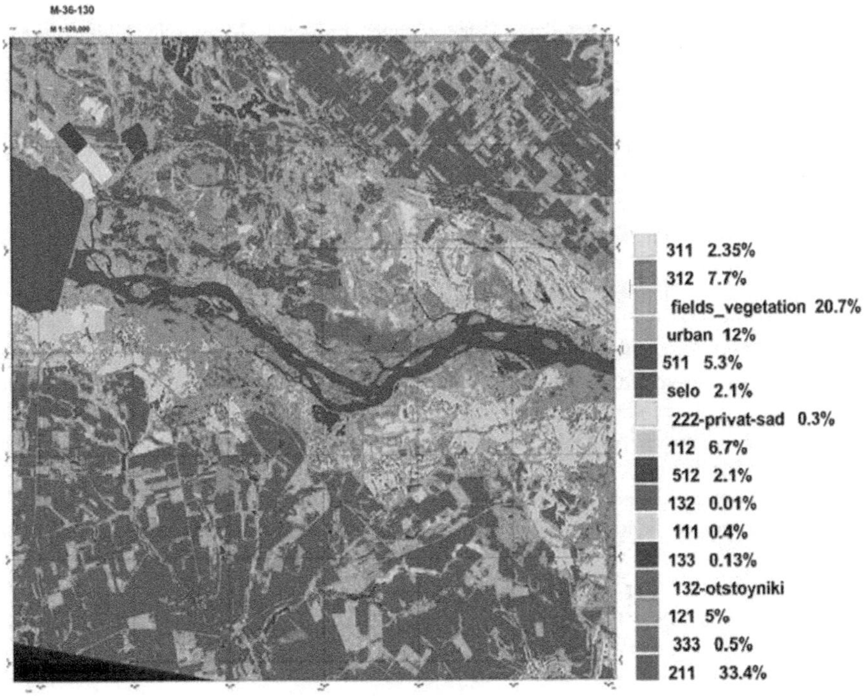

Figure 6. Classification result, M 1:100000

Multilevel (or multistage) model of classification procedure.

Post-classification model.

Validation method of classification results. Validation results for some test sites located in different landscape zones.

Development Software Modules for new models and algorithms. It is envisaged to use ENVI & IDL and eCognition as technological platform.

To successful solving of the problems put by the Dniprocosmos State Company is interested in collaboration with institutions, which have the necessary experience and knowledge in thematic processing of Earth remote sensing data.

References

Building an Information Capacity for Environmental Protection and Security. A contribution to the initial period of the GMES Action Plan (2002–2003). European Commission, Directorate-General for research sustainable development, global change and ecosystems, EUR 211109.

G. Büttner, The European CORINE land cover database, ISPRS Commission VII Symposium, Budapest, September 1–4, 1998. Proceedings, 633–638.

EEA-ETC/TE, 2002. CORINE land cover update. I&CLC2000 project, Technical Guidelines, http://terrestrial.eionet.eu.int.

INSPIRE Architecture and Standards Position Paper. European Commission, Joint Research Centre, 2002, EUR 20518 EN, – INSPIRE AST PP v4–3 en.doc.

INSPIRE – work programme preparatory phase 2005–2006. Final draft. September 05, 2004, ESTAT-JRC-ENV.

V. I. Lyalko and M. O. Popov (eds.), Multispectral Remote Sensing in Nature Management, Kyiv, Naukova Dumka, 2006, http://www.nkau.gov.ua.

The role of remote sensing of the earth in formation of the national infrastructure of an environment spatial monitoring data – V. I. Voloshyn, Y. I. Bushuyev, A. G. Shapar, O. P. Fedorov – Ecology and Nature Management: Collection of Scientific Works of Institute of Problems on Nature Management and Ecology of the NAS of Ukraine, Issue 8, Dniepropetrovsk, 2005a, pp. 141–151 (in Russian).

V. I. Voloshyn, V. M. Korchinsky, M. M. Kharytonov, and M. K. Sundareshan, A novel method for correction of distortions and improvement of information content in satellite-acquired multispectral images, Proceedings of NATO ASI on MultisensorData and Information Processing for Rapid and Reliable Situation and Threat Assessment, Albena, Bulgaria, May 2005.

V. I. Voloshyn, Y. I. Bushuyev, O. I. Parshina, and O. P. Fedorov, Method of classification of integumentary landscape elements, Space Science and Technology, 10(5/6), 2004 (in Ukrainian).

PRODEC – EMERGENCY PROCEDURE BASED ON FUZZY NOTIONS FOR CATCHMENT MANAGEMENT

JAN W. OWSIŃSKI, ANDRZEJ ZIÓŁKOWSKI
Polish Academy of Sciences, Institute of Fundamental Technological Research, Warsaw, Poland
aziolk@ippt.gov.pl

Abstract. The paper describes the ProDec system serving to develop and run decision procedures based upon IF...THEN rules. The system encompasses the editing, testing, and running functions. The key feature consists in the use of fuzzy values of the conditioning parameters. These fuzzy values are defined in the development and editing stage and can be easily referred to in the testing and use stages through natural language expressions. In this manner the inherent uncertainty associated with virtually all decision situations is captured and communication with the procedure is facilitated. The use of ProDec for river catchment emergency management purposes is illustrated.

Keywords: decision procedure, IF...THEN... rules, emergency management, fuzzy values, natural language expressions

1. Introduction

Taking of correct decisions in the emergency situations, such as flood or spilling of dangerous chemicals may save human life and property. Taking of such decisions occurs under severe stress. Decisions have to be taken quickly, and the potential mistakes may have catastrophic consequences. That is why it is essential to prepare beforehand the principles of proceeding for the cases of all the potential threats in order to be able to properly react in every such situation. It is necessary to train future decision-makers, so as to provide them with knowledge on the ways of reacting to hazards. This requires to carry out preparatory analyses and elaborate the responses to threats, due account being taken of the technical and financial possibilities.

A. Morris and S. Kokhan (eds.), Geographic Uncertainty in Environmental Security, 271–283.

Earlier preparations are often neglected, because they require bearing some cost. Yet, this cost is incomparable with the losses that can arise by neglecting the preparations.

The ProDec (*Pro*cedures for *Dec*isions) software has been elaborated within the framework of the EU TransCat Project, whose purpose was to develop the system for supporting decisions in transboundary catchment management.

TransCat was a 5FP project that ended in January 2006, meant to produce an integrated web-based DSS for transboundary river catchment management. Such a system was indeed developed (transcat.vsb.cz, transact.ibspan.waw.pl), centered on data management core, emphasizing mapping functionality. Polish team on the project was responsible for the development of a range of decision-oriented applications that were meant to serve in various kinds of decision situations (two-sided option evaluation, group decision-making, public participation, etc.). ProDec made a part of this set of applications.

2. The Principles of Functioning of the ProDec System

The ProDec system makes use of simple decision rules of the following type:

If a definite threat takes place
then undertake a definite action

The states of threat or hazard are recognized on the basis of the current values of the observed indicators. In order to state whether an emergency state actually takes place, we have to check whether the indicator values are contained in definite intervals. So, on the basis of known values of indicators we check what *kinds* or *degrees* of threats may occur, and then, by applying the decision rules we determine the list of actions corresponding to these kinds or degrees, that is, to the state identified.

The novelty in the ProDec system is constituted by the use of fuzzy logic to define the decision rules, and more precisely – to define the degrees or states of threat or emergency on the basis of the indicators observed. The precise intervals of values, which define the degree of emergency, have been replaced by the intervals with fuzzy border. This is equivalent to fuzzy ("linguistic") definition of the hazard-related notions based on indicator values.

In ProDec a complex decision tree can be defined as a collection of simple rules. The first rule is the starting point of the decision tree design and analysis. Each rule is a collection of:

- Conditional expressions
- Actions
- References to other rules (that is – further conditions and actions)

In fact, any rule is a decision subtree, whereas the first rule represents the whole tree. The decision tree can be defined as a single, even if quite complex, rule, but it might be convenient to define more rules if the same subtree appears more than once. The typical decision tree structure with few rules is shown in Figure 1.

The Conditional expression (IF *state* THEN *actions*) has two parts:

Condition
Action part

For example, if it is raining, then open the umbrella. Such conditional expression defines actions that should be taken if some state occurs. Each condition is described in ProDec by a single state name but the state entering the condition may be composite. It is defined as a simple fuzzy logical expression given in a readable, linguistic form:

IF water level is very high OR water level is high AND the rain forecast is high THEN ...

The text "water level is very high" in the above example represents the fuzzy value "very high" related to the "water level" parameter. All fuzzy values are described by four parameters necessary to define the trapezoidal membership function (Figure 2).

The membership function of the fuzzy water level value illustrated in Figure 2 ("high water level") equals 1 for the range 315–360 of water levels. Any specific value from this range is classified as "high water level".

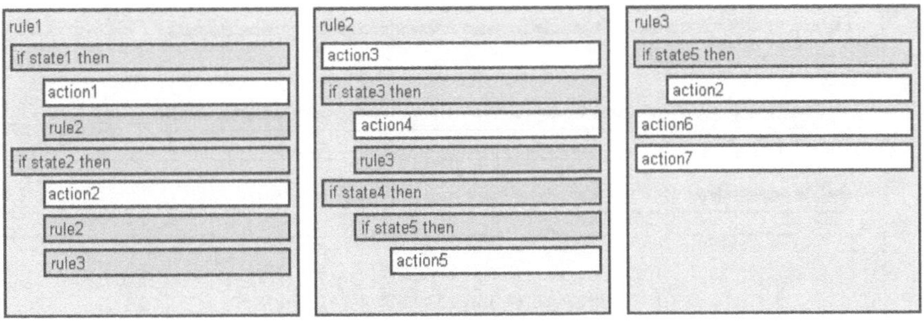

Figure 1. Decision tree structure

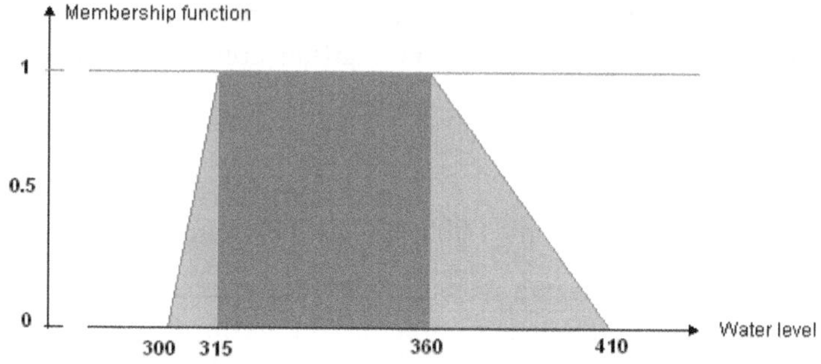

Figure 2. Membership function

Values below 300 and above 410 are definitely not "high water levels" (in the latter case they might belong to the value of "very high water level"). Water levels in the ranges 300–315 and 360–410 belong to the fuzzy border of the "high water level" value, and they may overlap with borders of other fuzzy values.

In general, a fuzzy value represents some range of values, rather than an exact value, with graded "memberships" in the fuzzy value. Linguistic equivalents of fuzzy values are used in natural language, and so a decision tree involving fuzzy values can be easily understood. Figure 3 shows an example of definition of verbally designated water level "states" involving the use of "fuzzy" intervals of values.

By employing fuzzy logic we gain additional information in doubtful situations, especially when the current values of the indicators are close to

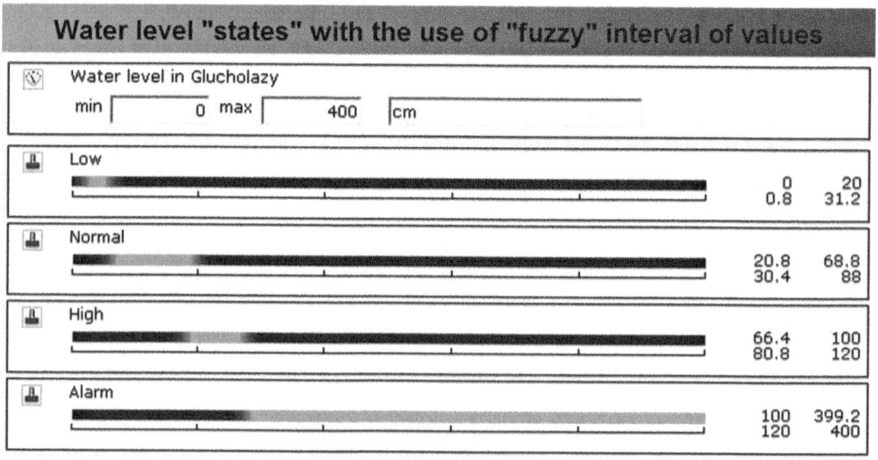

Figure 3. An example of state definition

the limits of the intervals indicating a proper emergency state. We obtain, as the result of application of the fuzzy decision rules, the list of actions together with their degrees of belongingness to ("membership in") the set of actions to be actually implemented. If the degree of belongingness or membership in this set is equal 1, then this action should unconditionally be carried out. The closer the membership degree to 1, the stronger the indication to implement a given action. The closer the membership degree to 0, the lesser the need for carrying out a given action. In addition, we are able to compare the membership degrees, or the strengths of indication for various potential actions (between "do nothing" and "raise the red alarm").

3. Functions and Components of the ProDec System

The ProDec system is composed of two parts:

- The part supporting the determination of the decision rules
- The part supporting decision making

In the first part the set of decision rules is prepared. On the basis of the measured or forecasted values of indicators, like, e.g., water level or forecasted precipitation, the rules determine whether emergency situation would arise. For various degrees of emergency and kinds of states the rules determine different actions to be undertaken. So, for instance, in case of the threat of local flooding, assistance ought to be secured in salvaging of property, while in case of indication of a large flood it will be necessary to carry out quick evacuation of the population.

The second part functions on the basis of the set of decision rules prepared with the use of the ProDec's first part. Knowing the current values of the indicators used to determine the degree of emergency, we apply these decision rules, and obtain the set of suggested actions, to be undertaken (or considered) in a given situation.

4. Designing Decision Procedure with the ProDec System

Preparation of the decision support system, whose functioning is based upon the decision rules, can be divided into the stages specified below. These stages are listed here in the sequence of their execution, but in the course of work there may arise the need of complementing and adding precision to the information prepared in an earlier stage.

4.1. DETERMINATION OF THE THREATS AND ACTIONS

The very first step in the elaboration of decision rules is to establish the list of all the possible threats and degrees of emergency, as well as actions, which might be undertaken in order to deal with the threats. It is important to give the threats and actions short, but unambiguous names, because these names will be used in the definition of the decision rules. If we envisage undertaking of different actions depending upon various degrees of emergency, we should treat these degrees as different "kinds of threat".

An example of threats that can be defined for a river valley is as follows:

- Small local flooding event
- Larger local flooding without threat to human life
- Serious flood
- Effluence of dangerous chemical substances from a cistern

And the examples of actions:

- Announcing the state of emergency
- Strengthening of the flood protection walls
- Emptying of reservoirs
- Announcement of flood alarm
- Evacuation of population from the threatened areas
- Supply of drinking water to the flooded areas
- Supplying the flooded population with food

4.2. DIVISION INTO SUBAREAS

In order to correctly elaborate the decision rules we must divide the area under threat into the subareas featuring (potentially) similar degrees and character of hazard. In case of floods this division results mostly from terrain relief, but it may also be associated with the nature of land use, administrative breakdown, etc.

A too detailed division requires definition of a high number of decision rules, and so we should distinguish only such subareas, over which the hazards are clearly different, and the ones, which require different actions in the case of emergency. Thus, for instance, we can distinguish the areas situated close to the river channel within the floodplain, which are often flooded, and the areas, which are flooded only sporadically. On the op of this, we can determine the urban areas in view of the necessity of taking additional actions (Figure 4).

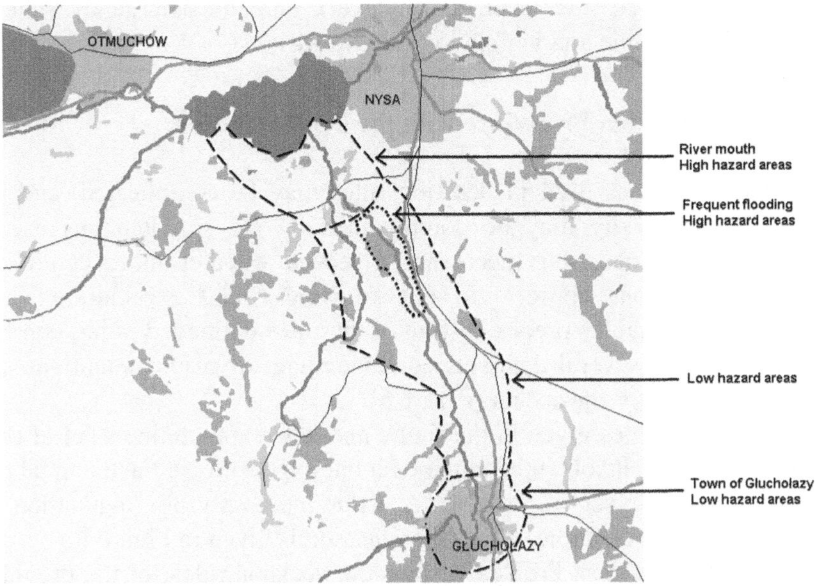

Figure 4. Division into subareas

4.3. THE CHOICE OF MEASURED QUANTITIES

Rational decisions can be taken only on the basis of available information. When designing a decision-making supporting system we have to choose the quantities to constitute the basis for decision taking. Obviously, information on these quantities has to be always available and up to date, since otherwise our decisions risk to be completely erroneous. While defining the decision rules, we will often perceive the need of organizing or complementing the system of monitoring the emergency situations, which would regularly provide information needed for decision taking.

For purposes of determination of the degree of flood threat over a given area we can use the measurements of water level and flow upstream from the area, as well as the current forecasts of precipitation.

4.4. DEFINITION OF THE DEGREES OF EMERGENCY

In order to conclude whether we actually deal with flood emergency we have to know the current values of the indicators selected in the preceding step as the basis for taking of decisions. Depending upon the locations of values of these indicators in terms of appropriate value intervals we will conclude whether we deal with actual emergency or not. In the ProDec system we assign the ranges of indicator values telling names (e.g., high level, normal level, low level, alarm level), and the emergency state is defined with a simple and legible logical expression.

5. Testing Decision Procedure with the ProDec System

Since the conditions used in decision rules may be complicated, and the degree of complexity may be compounded by the simultaneous use of several fuzzy values, it is necessary to check whether these conditions indicate the actions conform to our knowledge and expectations. The ProDec system enables precise testing of the rules defined. Testing consists in preparation of several data sets corresponding to various situations and running the rules for these data (Figure 5).

For decision rules given in the figure above we may define a set of tests to check if the results obtained agree with our expectations. Each named test is defined by the set of parameters value necessary for evaluation of decision rules. An example of test specification is given in Figure 6.

For the selected test ProDec verifies the decision rules for the provided values of indicators and shows the results indicating the “necessity” grades assigned the actions considered (Figure 7). For purposes of testing we can obtain a more detailed information, showing how the final result was generated.

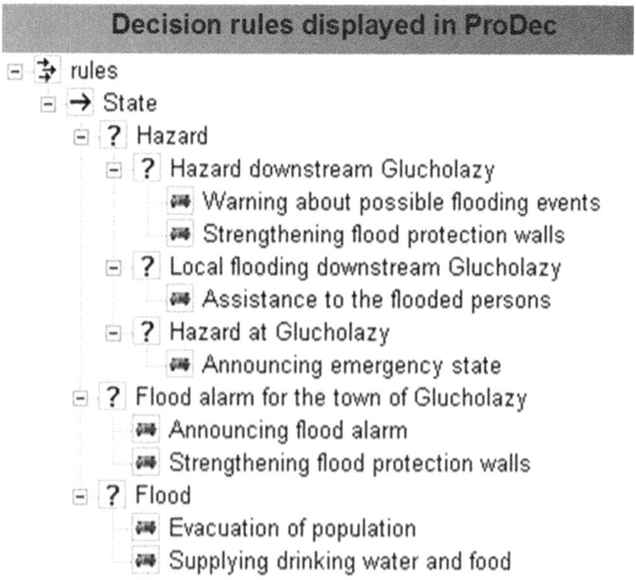

Figure 5. Decision rules

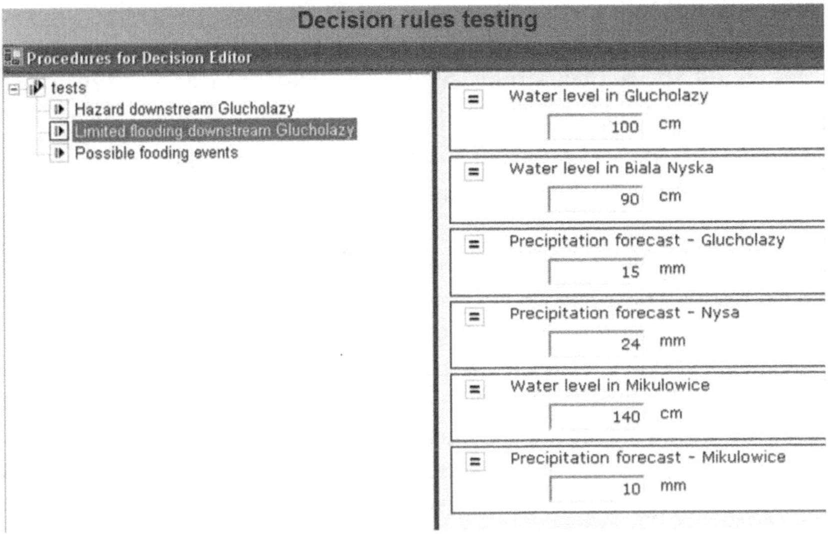

Figure 6. Decision rule testing: test specification

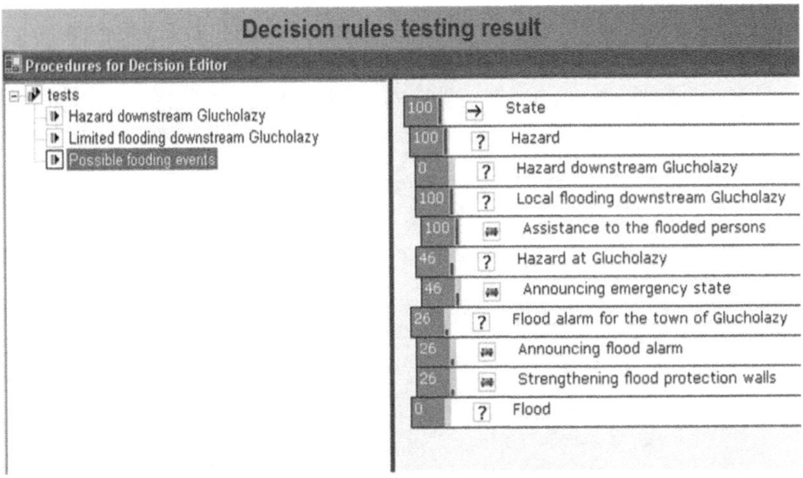

Figure 7. Decision rule testing: results obtained

The rows with the question marks correspond to the verified conditions (states). Number 0 against the gray background in the rows of this type means that upon the verification of conditions we have concluded that this state does not take place and none of the corresponding actions must be undertaken. The number 34 against the gray background in the row suggesting "Strengthening of the flood protection walls" defines the degree of membership of the respective set of data to the state "Threat downstream of Głuchołazy" on the scale between 0 and 100, meaning that the suggestion (of "necessity") of carrying out this action is rather weak. If this value were 85 we could speak of a strong suggestion, and for 100 – of an absolute necessity of executing this action.

6. Implementing Decision Procedure

Implementation of decision procedure developed with the help of the ProDec system is usually very simple. We need to measure and deliver actual values of parameters used in decision procedure. These values could be taken from some database or input by the user interactively. Some of parameters could be calculated as output of some complex models, e.g., a groundwater model. Having actual values of parameters and applying designed decision rules we will obtain the list of actions with indicators showing the degree of "necessity" of undertaking. An example of the interface used in ProDec system is given in Figure 8.

The intention was not to create a system that will substitute decision-makers. In some situations fully automated decision-making system could be advantageous, e.g., when there is a lack of time for a human decision maker to do the job or when correct decision is based on complex calculation.

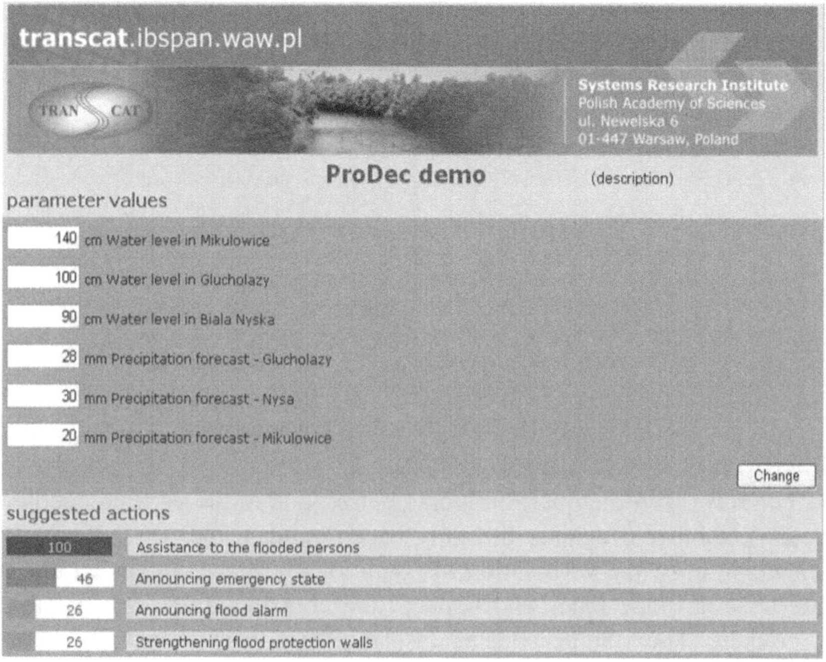

Figure 8. ProDec demo

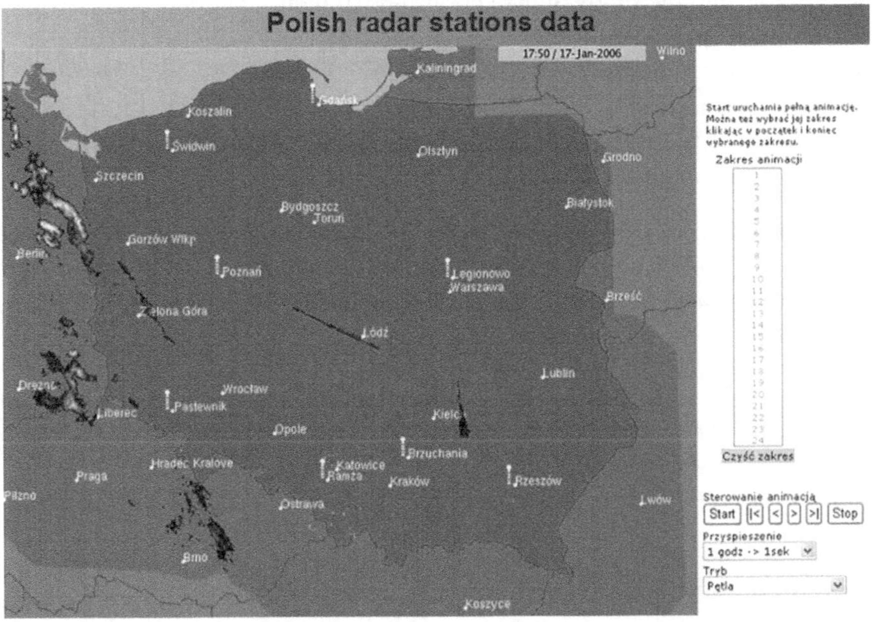

Figure 9. Radar station data

In such cases, however, fuzzy decision rules and fuzzy values are much less justified. Automated decision system must offer final decision without any space for human corrections. Implementation of fuzzy decision rules shows its advantages in advisory type of decision support system. Depending on measured parameter values we may obtain either precise advice which actions are best in this situation or a imprecise advice which may suggest alternative solutions along with their characterizations.

Because the results from ProDec are rather "suggestions" or "advices" as to which of actions should be taken, it is quite natural that we need to support decision-maker with additional sources of information to help him to make the final decision, but also to feed the values into the procedure. For example, we may support decision-maker with a current view of radar station data (Figure 9).

Data from the radar station, displayed in a graphical form, could not be easily transferred to parameter values (so it is difficult to use it directly as an input), but it may be analyzed by a decision-maker and may be helpful in the decision-making process.

The final stage of preparation of the ProDec system to practical use is elaboration of the detailed descriptions of all the actions included. These descriptions ought to contain the following information:

- The goal and the scope of the actions undertaken
- Persons responsible for their execution
- Contact information
- Resources needed
- Potential difficulties that can be encountered during execution and the ways of overcoming them

Preparation of this information in the electronic form (possibly as web pages in HTML language) offers additional advantages. This information can then be easily accessed, found, and updated. Some examples of action description are given in Figures 10 and 11.

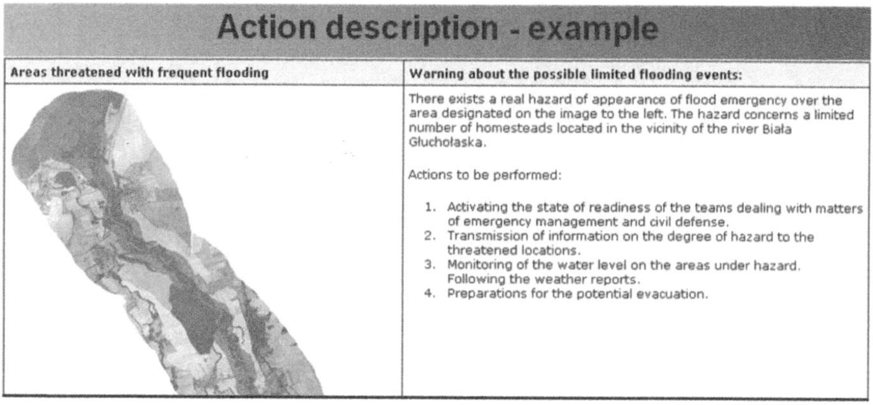

Figure 10. Action example

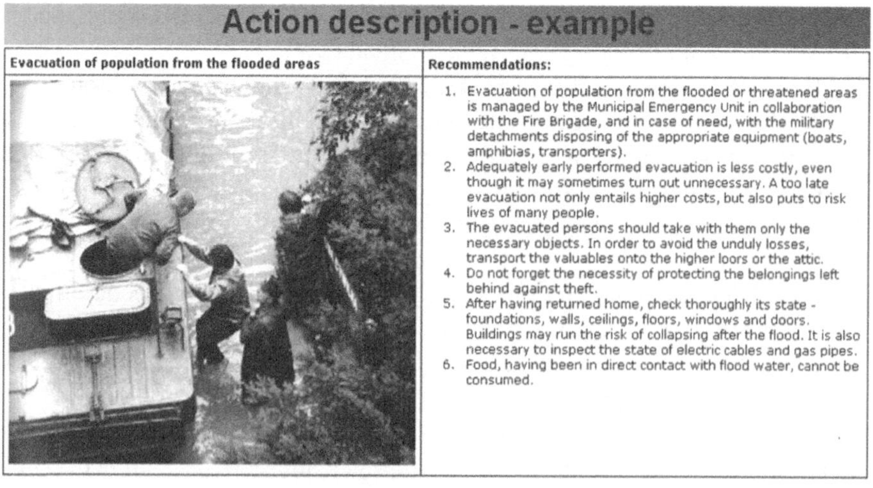

Figure 11. Action example

7. Concluding Remarks

The use of fuzzy values of conditioning parameters in the ProDec system allows for the formulation of the decision rules as legible phrases similar to those uttered in the natural language. Consequently, it is easier to understand and analyze the principles of decision taking and improve the rules adopted.

Reference to fuzzy values, corresponding to natural language expressions, in ProDec, was justified in the TransCat project by the assumption of functioning in "poor" conditions, that is – inadequate measurement and communication capacities, which have to be complemented by human experience, knowledge, and estimates, or guesstimates. Yet, applicability of the system extends well beyond such situations. Even if we use supposedly precise measurements and models, the entire system of interrelations is complex enough, always containing knowledge gaps, to warrant a special place for human experience and intuition in specifying parameter values corresponding to definite, potentially hazardous outcomes. These values and related conditions shall as a rule take a fuzzy form, corresponding to human language expressions.

No doubt, elaboration of the decision rules constitutes a substantial effort. Yet, even undertaking of such effort can pay back by itself. Frequently during such analysis various shortcomings are uncovered and proposals of new solutions emerge.

It is important to secure participation of several persons in the process of elaboration of decision rules in order to be able to confront various opinions and assessments concerning the evaluation of hazards and the best counteractions. Writing down the decision rules in a formalized manner and elaboration of the complementary documentation in the form of files accessible through internet facilitates communication and identification of various kinds of gaps and errors in the preparations.

The basic concept behind ProDec has given rise to some further projects, oriented at decision-making in different fields of human activity.

INDEX

A

ambiguity, 102
approximation, 122
artificial intelligence, 35
Association rule mining, 207

B

Bayesian, 246
Boolean logic, 134

C

consensual fusion, 150
Convex hull, 8
Crisp regions, 6

D

data cubes, 207
data warehouses, 209
decision rules, 280
Decision support systems, 201
DECORANA, 58, 61, 63
Dempster-Shafer, 244, 245
digital elevation model, 38

E

ecogeographic variables, 36
ecological evaluation, 156
economic development, 250
economics, 254
ecotone, 20, 22, 24, 26, 28
ecotones, 19, 21
egg yolk, 2, 4
environmental security, 256
error, 92
Error Assessment, 92
Errors, 123

F

feature pixels, 231
fitness for use, 114
floods, 276
fuzzy association rules, 201
fuzzy cardinality, 6
fuzzy cellular automata, 41
Fuzzy Certainty Measure, 89, 90, 103
Fuzzy change analysis, 167
fuzzy change matrix, 171
fuzzy classification, 37, 54, 62
Fuzzy classification, 172
Fuzzy clustering, 45, 47
fuzzy clusters, 55
fuzzy c-means, 40, 47, 58, 61, 69, 169
fuzzy confusion matrices, 100
fuzzy data cubes, 204
fuzzy data mining, 202
fuzzy decision rules, 275, 281
Fuzzy *k*-means, 82
fuzzy logic, 19, 38, 134
fuzzy numbers, 15
fuzzy OLAP, 220
fuzzy queries, 15
fuzzy region, 4, 7, 9, 12, 14, *See*
fuzzy regions, 2, 3, 4, 6
Fuzzy rules, 145
fuzzy set, 3, 5, 10, 20, 167
Fuzzy set, 168
fuzzy sets, 34, 41, 45, 47, 137
Fuzzy Sets, 36, 244, *See*
fuzzy spatial data cube, 217, 221
Fuzzy Spatial Data Cube, 210
fuzzy spatial data cubes, 216
fuzzy surface area, 6
fuzzy type 2 sets, 19

G

geometric point of view, 96

H

habitat, 40, 41, 44, 46
Habitat Suitability Index, 38
hard classifiers, 242
Human Development Index, 253

I

image classification, 242
indeterminate boundaries, 96
indiscernibility relation, 75
information causation, 113
Interactive segmentation, 226

L

Land cover mapping, 167
linguistic variables, 37

M

Markov chain, 113
Maximum Likelihood, 244
membership, 3, 4, 15, 21, 22, 28, 55, 63, 99, 101, 136, 158, 167, 171, 182, 242, 275
membership', 53
memberships, 173
minimum bounding rectangle, 4, 7
minimum bounding rectangles, 187
Monte Carlo, 83
multidimensional databases, 202

N

National Space Agency of Ukraine, 261
natural language, 123

O

OLAP, 202, 204
Ontology, 107

P

Pareto-*dominance*, 80
phrygana, 53, 72
Phrygana, 59
pixel force field, 225
ProDec, 271

Q

Quality of information, 108

R

remote sensing, 259, 260
Remote sensing, 239
river catchment management, 272
Rough Set, 77
Rough Sets, 75
RS. *See* Rough Sets
R-tree, 188, 192
R-Tree, 192
R-trees, 187

S

Semantic Import Model, 167
semantic loop, 111
semantic point of view, 96
Similarity Relation Model, 167
soft classifiers, 244
Soft classifiers, 242
SOFT constraints, 143
spatial data cubes, 202
Spatially explicit population models, 33
surface water monitoring, 153
Sustainable Development, 249, 250, 252, 254

T

taxonomies, 91
topology, 8
Triangulated irregular networks, 12
TWINSPAN, 53, 55, 61
type 2 fuzzy set, 170

type 2 fuzzy sets, 20, 26, 167, 169, 171, 184
type *n* fuzzy sets, 169
type-2 fuzzy sets, 47

U

uncertainty, 1, 3, 15, 23 26, 33, 36, 38, 41, 45, 46, 47, 54, 62, 68, 70, 76, 79, 89, 90, 92, 94, 96, 97, 99, 102, 103, 107, 110, 115, 119, 121, 123, 124, 125, 127, 133, 135, 167, 169, 184, 187, 190, 244, 245, 255, 256, 271

V

Vague Spatial Boundaries, 188
vagueness, 100
Vegetation land cover, 245
Vinnytsia, 160
Vinnytsia Region, 153

W

water quality, 158
α-cut, 5, 168
α*MBR-cuts*, 192

GPSR Compliance

The European Union's (EU) General Product Safety Regulation (GPSR) is a set of rules that requires consumer products to be safe and our obligations to ensure this.

If you have any concerns about our products, you can contact us on ProductSafety@springernature.com

In case Publisher is established outside the EU, the EU authorized representative is:

Springer Nature Customer Service Center GmbH
Europaplatz 3
69115 Heidelberg, Germany

Batch number: 10809441

Printed by Printforce, the Netherlands